The Creation of the Universe

REVISED EDITION

OTHER BOOKS BY GEORGE GAMOW

ONE TWO THREE . . . INFINITY

PUZZLE-MATH (with Marvin Stern)

A PLANET CALLED EARTH

A STAR CALLED THE SUN

The Great Spiral Nebula in Andromeda

The Creation of the Universe

REVISED EDITION

GEORGE GAMOW

PROFESSOR OF PHYSICS
UNIVERSITY OF COLORADO

Published by The Viking Press · New York

FIRST PUBLISHED IN 1952 BY THE VIKING PRESS, INC.
THIRD PRINTING 1955

REVISED EDITION
PUBLISHED IN 1961 BY THE VIKING PRESS, INC.
625 MADISON AVENUE, NEW YORK, N.Y. 10022
FOURTH PRINTING NOVEMBER 1966

PUBLISHED SIMULTANEOUSLY IN CANADA BY
THE MACMILLAN COMPANY OF CANADA LIMITED

LIBRARY OF CONGRESS CATALOG CARD NUMBER: 61-7387
PRINTED IN THE U.S.A. BY VAIL-BALLOU PRESS, INC.

To fellow cosmogonists
of all lands and ages

FROM *The Creation* by Joseph Haydn

Preface to the Revised Edition

Since this book was written (almost ten years ago) a heated argument has developed between the proponents of the conventional theory of the Expanding Universe, now known as the "Big Bang" hypothesis, and those of the "Steady State" hypothesis of Gold, Bondi, and Hoyle. According to the latter theory the universe has existed in a more or less unchangeable state through the eternity of the past, and will continue to exist in the same state for the eternity to come. The pro and contra arguments for both hypotheses are included in the present text of this book.

There were also new developments in the theory of the origin of chemical elements, and it seems now that whereas basic cooking of atomic nuclei took place before stars were formed, an additional cooking of heavy elements must have taken place at a later date in the hot interiors of stars. These developments have also been incorporated into the new text.

In view of the objections raised by some reviewers concerning the use of the word "creation," it should be explained that the author understands this term, not in the sense of "making something out of nothing," but rather as "making something shapely out of shapelessness," as, for example, in the phrase "the latest creation of Parisian fashion."

G. GAMOW
University of Colorado

August 1960

Contents

Illustrations

PLATES

All photographs courtesy of Mount Wilson and Palomar Observatories

FIGURES

The Creation of the Universe

Introduction

"Give me matter and I will construct a world out of it."
—IMMANUEL KANT,
Allgemeine Naturgeschichte und Theorie des Himmels

The problems of cosmogony—that is, the theory of the origin of the world—have perplexed the human mind ever since the dawn of history. Among the ancients, the origin of the world was necessarily associated with a creative act by some deity, who separated light from darkness, raised and fixed the heavens high above the surface of the earth, and fashioned all the other features that characterized the highly limited world picture of early man.

As the centuries rolled by and men gradually accumulated knowledge about the various phenomena taking place in the world that formed their environment, the theories of cosmogony took a more scientific shape. The names of Buffon, Kant, and Laplace characterize the scientific era when the first attempts were made to understand the origin of the world exclusively as the result of natural causes. The theories of that time, which were limited essentially to the origin of our solar system, later underwent a process of multiple evolution; culminating in a reasonably complete and consistent theory of planetary formation recently developed by Carl von Weizsäcker and Gerard P. Kuiper.

In the meantime the progress of observational astronomy opened entirely new horizons of knowledge of the universe and reduced the old riddle of the birth of planets to a minor incident within a much broader picture of the evolution of the universe. The main problem of cosmogony today is to explain the origin and evolution of the giant stellar families, known as galaxies, which are scattered through the vast expanses of the universe as far as can be seen with the strongest telescopes. The key factor for the understanding of this large-scale cosmic evolution was provided about a quarter-century ago by a discovery of the American astronomer Edwin P. Hubble. Hubble found that the galaxies populating the space of the universe are in a state of rapid dispersion ("expanding universe"). This implies that once upon a time all the matter of the universe must have been uniformly squeezed into a continuous mass of hot gas. The close correlation between the observed phenomenon of expansion and certain mathematical consequences of Einstein's general theory of relativity was first recognized by an imaginative Belgian scientist, Abbé Georges Edouard Lemaître, who formulated an ambitious program for explaining the highly complex structure of the universe known to us today as the result of successive stages of differentiation which must have taken place as a concomitant of the expansion of the originally homogeneous primordial material. If and when such a program is carried through in all details, we shall have a complete system of cosmogony that will satisfy the principal aim of science by reducing the observed complexity of natural phenomena to the smallest possible number of initial assumptions. Although such a program is far from completion as of today, considerable progress has been made on various parts of it, and the end seems to be already in sight.

It must be remarked here that at present there still exist rather fundamental differences between the points of view accepted by various scientists working in this field. Many of them (including the author of the present book) believe that the present state of the universe resulted from a continuous evolutionary process, which started in a highly compressed homogeneous material a few billion years ago—the hypothesis of "beginning." Others prefer to consider the universe as existing in about the same state throughout eternity—the hypothesis of a "steady-state universe." One of the proponents of the latter view in the field of stellar evolution is the noted Russian astronomer Vorontzoff-Velyaminov,[1] who was apparently forced by the philosophy of dialectic materialism to accept this hypothesis. In a rather different form, and certainly for an entirely different reason, similar views are held by the British astronomer Fred Hoyle,[2] who attempts to explain the alleged steady state of the universe by introducing a hypothesis of continuous creation of matter in intergalactic space.

It is probably too early to say which of the two points of view will ultimately prove to be correct. The main purpose of this book is to present the arguments in favor of the hypothesis of a "beginning" and to analyze critically the claims of the proponents of a steady-state universe.

It is hoped that this book will constitute an adequate survey of the subject for scientists in various fields, and at the same time be of service to laymen interested in the problems of modern cosmogony.

[1] A. Vorontzoff-Velyaminov, *Gaseous Nebulae and New Stars* (in Russian), Academy of Sciences, U.S.S.R., Moscow, 1948. Reviewed by O. Struve, *Astrophysical Journal*, 1949, pp. 110, 315.

[2] Fred Hoyle, *The Nature of the Universe*, Harper & Brothers, New York, 1951.

CHAPTER I

Evolution Versus Permanence

Before we can discuss the basic problem of the origin of our universe, we must ask ourselves whether such a discussion is necessary. Could it not be true that the universe has existed since eternity, changing slightly in one way or another in its minor features, but always remaining essentially the same as we know it today? The best way to answer this question is by collecting information about the probable age of various basic parts and features that characterize the present state of our universe.

The age of the atoms

For example, we may ask a physicist or a chemist: "How old are the atoms that form the material from which the universe is built?" Only half a century ago, before the discovery of radioactivity and its interpretation as the spontaneous decay of unstable atoms, such a question would not have made much sense. Atoms were considered to be basic indivisible particles and to have existed as such for an indefinite period of time. However, when the existence of natural radioactive elements was recognized, the situation became quite different. It became evident that if the atoms of the radioactive elements had been formed too far back in time, they would by now have decayed completely and disappeared. Thus the observed relative abundances of various radioactive elements may give us some clue as to the time of their origin. We notice first of all that thorium and the common isotope of uranium (U^{238}) are not markedly less abundant than the other heavy elements, such as, for example, bismuth, mercury, or gold. Since the half-life periods [1] of thorium and of common uranium are 14 billion [2] and 4.5 billion years, respectively, we must conclude that these atoms were formed not more than a few billion years ago. On the other hand, as everybody knows nowadays, the fissionable isotope of uranium (U^{235}) is very rare, constituting only 0.7 per cent of the main isotope; otherwise the Manhattan Project would have been as easy as fishing in a barrel. The half-life of U^{235} is considerably shorter than that of U^{238}, being only about 0.9 billion years. Since the amount of fissionable uranium has been cut in half every 0.9 billion years, it must have taken about seven such periods,[3] or about 6 billion years, to bring it down to its present rarity, if both isotopes were originally present in comparable amounts.

Similarly, in a few other radioactive elements, such as naturally radioactive potassium, the unstable isotopes are also always found in very small relative amounts. This suggests that these isotopes were reduced quite considerably by slow decay taking place over a period of a few billion years. Of course, there is no a priori reason for assuming that all the isotopes of a given element were originally produced in exactly equal amounts. But the coincidence of the results is significant, inasmuch as it indicates the approximate date of the formation of these nuclei. Furthermore, no radioactive elements with half-life periods shorter than a substantial portion of a billion years are found in nature, although they can be produced artificially in atomic piles. This also indicates that the formation of atomic species

[1] The half-life period of a radioactive substance is the period of time required to cut its original quantity by a factor of two. Thus after two, three, etc., half-life periods, only one-quarter, one-eighth, etc., of the original quantity will be left.

[2] The term "billion" is used to mean one thousand million.

[3] Because $(1/2)^7 = 1/128 = 0.8$ per cent.

must have taken place not much more recently than a few billion years before the present time. Thus there is a strong argument for assuming that radioactive atoms and, along with them, all other stable atoms were formed under some unusual circumstances which must have existed in the universe *a few billion years ago.*

The age of the rocks

As the next step in our inquiry, we may ask a geologist: "How old are the rocks that form the crust of our globe?" The age of various rocks—that is, the time that has elapsed since their solidification from the molten state—can be estimated with great precision by the so-called radioactive-clock method. This method, which was originally developed by Lord Rutherford, is based on the determination of the lead content in various radioactive minerals such as pitchblende and uraninite. The significant point is that the natural decay of radioactive materials results in the formation of the so-called radiogenic lead isotopes. The decay of thorium produces the lead isotope Pb^{208}, whereas the two isotopes of uranium produce Pb^{207} and Pb^{206}. These radiogenic lead isotopes differ from their companion Pb^{204}, natural lead, which is *not* the product of decay of any natural radioactive element.

As long as the rock material is in a molten state, as it is in the interior of the earth, various physical and chemical processes may separate the newly produced lead from the mother substance. However, after the material has become solid and ore has been formed, radiogenic lead remains at the place of its origin. The longer the time period after solidification of the rock, the larger the amount of lead deposited by any given amount of a radioactive substance. Therefore, if one measures the relative amounts of deposited radiogenic lead isotopes and the lead-producing radioactive substances (that is, the ratios: Pb^{208}/Th^{232}, Pb^{207}/U^{235}, and Pb^{206}/U^{238}) and if one knows the corresponding decay rates, one can get three independent (and usually coinciding) estimates of the time when a given radioactive ore was formed. By applying this method to radioactive deposits that belong to different geological eras, one gets results of the kind shown in the following table.

THE AGE OF VARIOUS RADIOACTIVE MINERALS

MINERAL	LOCALITY	GEOLOGICAL PERIOD	ESTIMATED AGE (MILLIONS OF YEARS)
Pitchblende	Colorado (U.S.)	Tertiary	58
Pitchblende	Bohemia (Europe)	Carboniferous	215
Pitchblende	Belgian Congo (Africa)	Pre-Cambrian	580
Pitchblende	Great Bear Lake (Canada)	Pre-Cambrian	1,330
Uranite	Karelia (U.S.S.R.)	Pre-Cambrian	1,765
Uranite	Manitoba (Canada)	Pre-Cambrian	1,985
Monazite	Rhodesia (Africa)	Pre-Cambrian	2,710

The last mineral in the table is the oldest yet found, and from its age we must conclude that the crust of the earth is at least 2.7 billion years old.

A much more elaborate method was proposed recently by the British geologist Arthur Holmes. This method goes beyond the formation time of different radioactive deposits and claims an accurate figure for the age of the material forming the earth. Perhaps the simplest way to illustrate it is by way of a story about an absent-minded Western rancher. This rancher remembers that one day in the spring he let all his cattle out to graze on his pastures, but he cannot recall the exact date on which he did so. He also remembers that at various dates during the summer he was collecting the cattle from different pastures and locking them into newly built corrals (one corral on each pasture), but these dates he has forgotten too. Is there any way for him to reconstruct the sequence?

Yes, there is, provided that he does not mind handling the dung produced by his cattle in the corrals and on the pastures. The reader has probably guessed that the dung produced by the cattle serves here as a symbol of the lead produced by decaying uranium, and that locking up the cattle in corrals represents the formation of radioactive deposits in solidifying rocks. It would be easy, of course, to find out at what approximate dates the different corrals were occupied by measuring the total amount of accumulated dung in each corral and then dividing that amount by the dung productivity of the corresponding herd. (This is exactly like the radioactive-clock method for determining the age of rocks.) But what about the date on which the cattle were first let out into the pastures—the date that radioactive atoms were formed?

At first glance it might seem possible to apply here a similar method, by collecting all the dung produced by the cattle while they were grazing in the open. However, this might not give a correct answer, since there could have been some "primordial" dung in the field which was present *before* the cattle were first let out (representing the lead which originated simultaneously with uranium during the epoch when all atoms were formed, and which was not produced by uranium decay at a later time). Of course, the same objection could be made against the use of the dung method for estimating the ages of the separate corrals; but there, because of the small area of the corrals, the amount of primordial dung can be readily discounted, compared with what the cattle would produce in the course of even a few days. In the open field, on the other hand, the situation is entirely different and the existence of primordial dung may influence the result quite appreciably.

Considering the problem in more detail, we must realize, assuming the amount of primordial dung on all pastures to be the same (hypothesis of the uniformity of the original atom-making), that there would be less dung in those pastures from which the cattle were driven into corrals at the earlier dates. (Indeed, geologists do find less radiogenic lead in rocks of higher geological age.) For each pasture we can start with that amount of dung which did not change from the time the cattle were put into the corral and go back in time, subtracting the day-by-day dung production. In this way we shall come to a day (in the spring) when the dung in the fields was at zero. If no primordial dung had been present, that day would represent the first date forgotten by the rancher. But if primordial dung were present (its amount, of course, being unknown), this procedure, applied to specific pastures, would fail to give us the answer. However, the situation is entirely different if we compare the data supplied by several pastures. The curves representing the past history of dung deposits on different pastures will be, generally speaking, different, since they depend on the size of the fields, the number of cattle, the dates of corral building, etc. But if we plot all these curves on one diagram (as in Fig. 1) they should intersect in the same point, giving us both the forgotten date on which the cattle

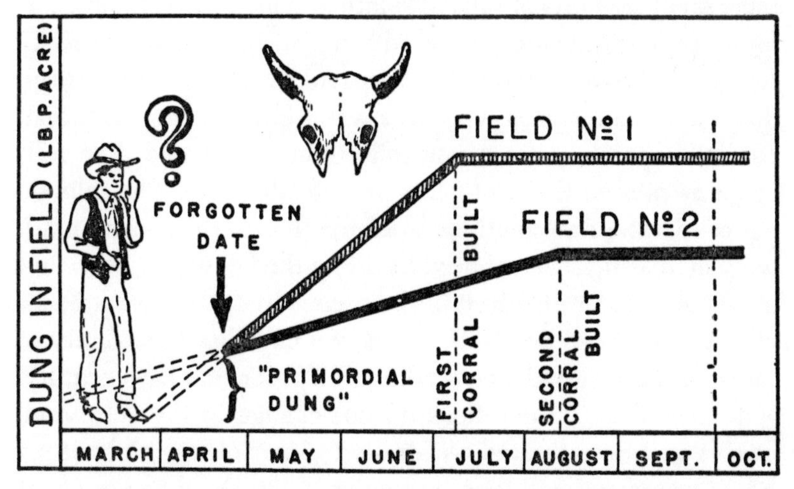

Fig. 1. How can one help a forgetful Western rancher find out when he let his cattle out to pasture? (Arthur Holmes' geological method)

were let out to pasture and the amount of primordial dung present at that date. By applying this method (amplified by the introduction of flocks of sheep and sheep dung, to account for the two uranium isotopes U^{238} and U^{235} and the two radiogenic lead isotopes Pb^{206} and Pb^{207}) to the relative amounts of lead isotopes found in rocks of different geological ages, Holmes found that all curves intersect near the point corresponding to a total age of 3.35 billion years,[4] which must represent the correct age of our earth. "Elementary, my dear Watson!"

The age of the oceans

Having received so much help from the geologists as regards solid deposits, let us turn to them once more with the question: "How old are the oceans that cover so much of the earth?" Here the answer is not quite so exact. The method was first proposed more than two centuries ago by Edmund Halley, who predicted the periodicity of the comet that bears his name. His method is based on the fact that the salinity of ocean water is mainly due to salts brought in by the rivers. We know that river water contains small amounts of salts in solution, which make it taste different from rain water. These salts are washed out of the rocky surface of the earth, mostly by fast rivulets and streams that run down mountain slopes. The river water brought into the ocean basins evaporates; the vapor collects in clouds and then falls again as rain on the continents in a steady cycle. The salts do not evaporate; they continue to accumulate in the oceans, gradually increasing their salinity. Dividing the known total amount of salt at present dissolved in the oceans by the known amount of salt brought in yearly by rivers, we find that the salinity of the oceans increases by one-millionth of 1 per cent each century. It follows that, if conditions do not change in the future, all oceans will be saturated with salts (36 per cent) in about 3.5 billion years and will then be similar to the Dead Sea or Great

[4] See Appendix, page 140.

Salt Lake. It also follows that the rivers must have been at work for about 300 million years for the present amount of oceanic salts (3 per cent) to have accumulated.

This figure, however, is on the short side, as the present rate of deposit of salt is known to be exceptionally high. The reason is that during the greater part of the history of our globe the surface of the continents was quite flat. The old mountains had been completely washed away into the oceans and new mountains had not yet been formed by the gradual contraction of the earth's crust. (Geologists count at least ten such successive mountain-raising periods.) It is estimated very roughly that during these flatland periods the erosive action of rivers must have been no more than one-tenth of what it is at present. This would result in an estimated age of our oceans of a few billion years, a figure which agrees with the estimated age of the oldest rocks.

The age of the moon

After thanking geology for all this valuable information, let us turn to astronomy and ask about the age of various celestial bodies, starting with the question: "How old is the moon?" We learn first that our Queen of the Night has not always been where it is now, that in the distant past it was so close to the earth one could almost touch it by stretching one's hand above one's head (if at that early epoch there had been any animals possessing hands and heads). As was shown by the work of the British astronomer George Darwin (son of the biologist Charles Darwin), the moon is constantly receding from the earth. Its distance from the earth increases at the rate of about 5 inches every year. It goes without saying that even the most precise instruments could not possibly measure such a slight increase in the distance to the moon, and that this conclusion was reached in a roundabout but nevertheless perfectly reliable way.

To understand the argument we must remember that the

interaction between the moon and the earth is most markedly displayed in the phenomenon of the tidal waves raised by the moon's attraction on the oceans of the earth. Tidal waves running around and around our globe encounter resistance in the form of the continents that stand in their way. If we could look at the earth-moon system from some fixed point in space, we should see the body of the earth rotating inside the two tidal bulges, much as the axle of a wheel rotates between two brake shoes. Thus we should expect that the rotation of the earth would be slowed down gradually and that this, in turn, would cause a gradual increase in the length of our day. According to a fundamental law of mechanics, known as the law of conservation of angular momentum, this lengthening of the day must result in a lengthening of the rotation period of the moon (month) and in a gradual increase in its distance from the earth.

It has been estimated that tidal friction will lengthen the day by about one-thousandth of a second per century, and will increase the length of the month by one-eighth of a second per century, besides causing the increase mentioned in the moon's distance from the earth. Small as they may seem, these estimated changes in the length of the day and of the month can be checked by direct astronomical observation. In fact, these changes advance the position of the sun among the fixed stars by 0.75 seconds of arc, and the position of the moon by 5.8 seconds of arc, every century. Actual observations give the values of 1.5 ± 0.3 and 4.3 ± 0.7, which are in reasonable agreement with the estimated effect. Consequently there can be little doubt about the accuracy of the estimated increase in the distance between the earth and the moon.[5]

[5] It is interesting to note that the lengthening of the day and of the month plays an important role in the correlation between the eclipses as reported by ancient writers and the calculations worked out by modern methods of celestial mechanics. Babylonian and Egyptian manuscripts dating as far back as the twentieth century B.C. give the dates of observed eclipses. These records usually give dates in perfect agree-

By means of mathematical calculations George Darwin reached the conclusion that the moon must have been practically in contact with the earth about four billion years ago. One surprising result of these calculations is that, at that time, the length of a month (moon's orbital period) was equal to the length of a day (earth's diurnal period), both being equal to 7 present-day hours!

At that early epoch, the moon must have hung motionless above the same point of the earth's surface, the point at which it was born by being drawn out from the mother body by the tidal forces of the sun. We may appropriately call this early state of our satellite a Hawaiian Moon, since in all likelihood its birthplace was the middle of the Pacific Ocean. In fact, there is evidence to support the assumption that the Pacific Basin is nothing but a giant scar in the granite skin of Mother Earth, a constant reminder of the birth of her first and only daughter.

The ages of the sun and other stars

But what about the sun and the stars—how old are they? We can estimate the present age of the Milky Way, the giant stellar family of which our sun is a humble member, by inquiring into the sources of energy which keep the stars hot and luminous. It

---

ment with the calculated dates, but disagree with the calculations with respect to the time of day when the eclipse was observed. In fact in some cases solar eclipses reported in the manuscripts would have taken place, according to calculation, several hours before sunrise or after sunset in the eastern Mediterranean so that they could not possibly have been observed in Babylon and Egypt! The explanation of the discrepancy lies in the fact that the theoretical calculations were made on the assumption that the length of the day is constant. However, if the day lengthens at the rate of 0.001 second every century, it must have been 0.04 seconds shorter 40 centuries ago. The average length of the day for this 40-century interval must be considered 0.02 seconds shorter than the present value. Now 40 centuries contain about 1,400,000 days, so that the total accumulated discrepancy is

$$1{,}400{,}000 \cdot 0.002 \text{ sec} = 28{,}000 \text{ sec} = 8 \text{ hr}$$

This is quite enough to explain the fact that the Babylonians and Egyptians observed these eclipses while the sun was above the horizon.

is an established fact that the energy generated in stars is caused by the gradual transformation of their original hydrogen content into helium. (This is discussed in more detail in Chapter V.) Nuclear transformation of hydrogen into helium is known to set free $2 \cdot 10^{-13}$ calories for each hydrogen atom utilized. Since our sun, for example, liberates 10^{26} calories per second, it must consume $5 \cdot 10^{38}$ atoms or about 800 million tons of hydrogen every second. On the other hand we also know that hydrogen constitutes about 50 per cent of the total mass of the sun, which is $2 \cdot 10^{27}$ tons. Thus it must take

$$\frac{1 \cdot 10^{27}}{8 \cdot 10^{8}} = 1.4 \cdot 10^{18}\ \text{sec} = 5 \cdot 10^{10}\ \text{yr}$$

to use up the whole hydrogen content. It seems, however, that the sun can use only about 20 per cent of its hydrogen, the amount which is contained in its central convective core. This lowers its total life span to only 10 billion years. But we must remember that stars of different sizes have different life spans, so that stars having the same chronological age may be in entirely different stages of their evolution. In fact, a small, a medium, and a large star, born at the same date in the past, may at the present time have reached different states of development, comparable to those reached by a year-old mouse, a year-old dog, and a year-old human being.

The reason for the different life spans of stars can be found in the fact that stellar brightness increases as the cube of the star's mass. Thus a star twice as massive as our sun uses up its fuel eight times as fast, and since the total amount of the fuel is only twice that of the sun (in proportion to total mass), the life span of such a star will be only one-fourth that of the sun. This difference in stellar life spans provides a convenient way to estimate the ages of the stellar population of our galaxy. Observational evidence indicates that a profound difference exists between stars which are lighter than four sun masses and those which are heavier than that. The stars of the first group form the main bulk of the stellar population and are very quiet members of that society. As soon as we cross the four sun-mass limit, we find that the number of stars is drastically reduced and those few stars which do fall into this category behave in a rather unusual way, many of them spinning wildly around their axes, ejecting streams of hot material from their equatorial bulges.[6] On the border line between these two main groups lie the stars which are apparently in a very unstable state and are up to all kinds of celestial tricks. Some of them swell up to enormous size and begin to pulsate periodically with change of brightness: the Cepheid variables. Others go through processes of outright explosion, ranging from minor flashes (periodic explosions of U Geminorum stars) to spectacular outbursts (supernovae) which make a single star temporarily as luminous as the entire galaxy to which it belongs.

In Chapter V there is more detailed discussion of these pulsations and explosions of stars which seem to be the symptoms of their old age, the last convulsions of stars which have almost used up their original fuel supply. Using the relationship previously mentioned between stellar masses and stellar life spans, we find that the mean age of the stars now approaching their thermal death is about 5 billion years. Thus we are led to the conclusion that most of the stars forming the Milky Way system were born about 5 billion years ago and that the few more massive stars observed in the sky are of more recent origin. As is more fully explained in Chapter V, the observed fast rotation of these more massive stars must be in part attributed to their comparative youth. As time goes on, and our stellar universe grows older and older, stars of lesser and lesser mass will gradually approach the end of their natural life and by the year 5,000,000,000 A.D. our own sun will get its death warrant.

[6] The stars of the first group are often found to rotate too, but their rotation, like that of our sun, is very slow.

The age of galactic clusters

Another method of estimating the age of the stellar population of our galaxy is based on the study of the purely mechanical behavior of stars forming the so-called galactic clusters; that is, the closely knit groups of stars that move together through the swarms of other stars in the Milky Way. One of these groups, found in the constellation Taurus, is shown in Fig. 2. The arrows

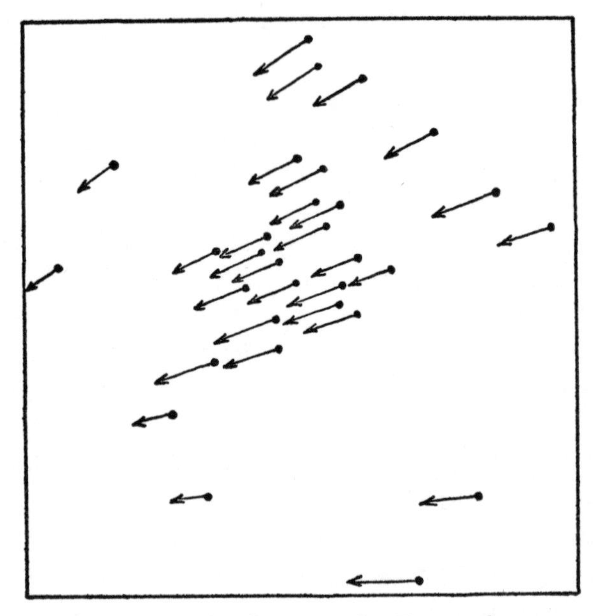

Fig. 2. Motion of stars in the Taurus cluster

indicate the displacement of the members of the group in the course of 500 centuries. We notice that the motion shows a "railroad track" effect, indicating that this particular group of stars is moving away from the present location of the sun. It is clear that such stellar groups, apparently formed by stars having a common origin from a single giant dust cloud, cannot hold together for an indefinite period of time. They must gradually be dissolved and scattered as the result of gravitational interaction with other stars that they encounter on their way. Calculations carried out by B. J. Bok at Harvard show that the average life of such stellar clusters must be between 1 billion and 10 billion years. Since there are still several hundred such clusters in our system, the Milky Way, its age cannot be greater than a few billion years.

The age of the Milky Way

A somewhat more general method of determining the age of the stars involves all the stars forming the Milky Way, and is based on the study of the distribution of energy among the stars. It is known that, apart from the regular whirlpool motion of the stars around the center of our galaxy, the stars also perform an irregular random motion, similar to the thermal motion of molecules in a gas. Owing to the gravitational interactions between the members of that giant swarm, the stars affect one another's motion and, after a certain period of time, are expected to arrive at a well-defined velocity distribution. According to the laws of statistical mechanics this final velocity distribution must correspond to the so-called equipartition of kinetic energy, the velocity of each star being inversely proportional to the square root of its mass.

These laws govern, in particular, the velocity distribution in a mixture of gases. Thus in a hydrogen-oxygen mixture the mean velocity of the molecules of oxygen is one-fourth that of the molecules of hydrogen, as their mass is sixteen times larger. However, while in an ordinary gas such equipartition of energy is established in a negligible fraction of a second, the similar process in a "gas of stars" takes a considerably longer time. In a recent work the German astronomer F. Gondolatsch has shown that at the present moment the equipartition of kinetic energy among the stars in the neighborhood of the sun is not yet completely established, but is about 2 per cent away from the final goal. That would mean, according to this theory, that the stellar system has been in existence between 2 billion and 5 billion years.

Thus we see that whenever we inquire about the age of some particular part or property of the universe we always get the same approximate answer: A FEW BILLION YEARS OLD.

It is true that the answers differ in respect to the exact number of billions, but they all agree as to the general order of magnitude. Thus it seems that we must reject the idea of a permanent unchangeable universe and must assume *that the basic features which characterize the universe as we know it today are the direct result of some evolutionary development which must have begun a few billion years ago.* We may also assume that in that distant past our universe was considerably less differentiated and complex than it is now and that the state of matter at that time could be accurately described by the classical concept of "primordial chaos." In fact, as the following chapters show, there is considerable empirical evidence pointing in this direction. With such an assumption, the problem of scientific cosmogony can be formulated as an attempt to reconstruct the evolutionary process which led from the simplicity of the early days of creation to the present immense complexity of the universe around us. In these inquiries we shall be greatly aided by the theory of the expanding universe which is discussed in Chapter II.

Plate I. Spiral nebula in Coma Berenices, a distant island universe seen on edge. Note the ring of darker matter encircling this nebula

Plate II. Spiral nebula in Canes Venatici, seen from the top. Note a satellite at the end of the lower arm

CHAPTER II

The Great Expansion

Broadening horizons

Astronomers have long been acquainted with certain rather peculiar-looking celestial objects which had come to be called "spiral nebulae." In contrast to other known nebulosities, which usually have irregular shapes and look more or less like clouds in the sky, spiral nebulae always have well-developed structures, consisting of a lenticular central body with a pair of spiral arms winding around it (Plates I and II). Until about a quarter of a century ago, the spiral nebulae were more or less generally believed to be located among the stars of our Milky Way system and were thought to be possible examples of stars giving birth to their own planetary systems according to the classical picture of Kant and Laplace.

However, in 1925, all these views were completely overthrown by a great discovery made by Edwin P. Hubble, an astronomer at Mount Wilson Observatory. Studying the Great Spiral Nebula of Andromeda, which is visually the largest of them all and therefore the most accessible to observation, he noticed that its spiral arms contain a number of extremely faint stars whose brightness changes periodically, following a simple sine law. Such stars, called "Cepheid variables" (after Delta Cephei, the first star in which such variability was noticed), are well known in our Milky Way system, and their periodic changes in luminosity are explained as the result of periodic pulsations of their giant bodies. A simple correlation exists between the period of these pulsations and the absolute luminosity of the star in question: the brighter the star the longer the period of pulsation. This so-called "period-luminosity relation" established by the Harvard astronomer Harlow Shapley is a powerful tool for measuring the distances of pulsating stars which are too far away to show a parallax displacement. By measuring directly the pulsation period of a given star, we can arrive at a definite conclusion about its absolute brightness. This, in combination with the visual brightness, tells us the actual distance of the star.

The observed pulsation periods of the Cepheids found by Hubble in the spiral arms of the Andromeda nebula indicated that they must possess very high absolute luminosities. On the other hand, they are so faint visually as to be at the limit of visibility. The inevitable conclusion was that they—and also the nebula itself—must be extremely far away. The distance to the Andromeda nebula estimated by that method worked out to almost 1 million light-years, that is, about a hundred times the diameter of the entire Milky Way system! Other spiral nebulae, which are visually smaller and fainter than the one in Andromeda, must be much farther away. If the spiral nebulae are really that far off, they must also be much larger than originally suspected; in fact they must be about as large as the Milky Way system itself!

Thus Hubble's discovery removed the spiral nebulae from their former humble position as common members of our galaxy and enthroned them as independent galaxies in their own right, scattered through the vast expanse of the universe. It became clear that these objects have nothing to do with ordinary nebulae (like the one in Orion), which are really only large clouds of dust floating in interstellar space. The spiral nebulae are formed by many billions of individual stars which blur together into one faintly luminous mass only because of their exceedingly great distance from the observer. More recently this conclusion was proved directly by another Mount Wilson astronomer, Walter Baade, who was able to resolve photographically the central body of the Andromeda nebula and those of its two companions into myriads of tiny luminous dots representing the individual stars from which these distant systems are built. It seems advisable to change the old terminology, and instead of talking about spiral nebulae to talk about spiral galaxies.[1]

Hubble's discovery proved to be the germ of still more remarkable progress in our knowledge about the nature of the universe. It had been known for some time that spectral lines in the light emitted by spiral nebulae show a shift toward the red end of the spectrum. Interpreted in terms of ordinary Doppler effect,[2] this meant that these objects were moving away from the observer. As long as these objects were believed to be members of our stellar system, one had to conclude that they had some peculiar motion among the stars, being driven from the central regions of the Milky Way toward its periphery. With the new broadening of horizons a completely new picture emerged: *the entire space of the universe, populated by billions of galaxies, is in a state of rapid expansion, with all its members flying away from one another at high speed.*

If the expansion of the space of the universe is uniform in all directions, an observer located in any one of the galaxies will see all other galaxies running away from him at velocities proportional to their distances from the observer. This can be easily demonstrated by gluing a number of pieces of paper (cut in the shape of galaxies, if desired) to the surface of a rubber balloon and blowing up the balloon to larger and larger size (Fig. 3). An observer located in *any one* of these model galaxies will see

[1] Some galaxies do not have spiral arms; these are called elliptical or spherical galaxies as the case may be.

[2] When the source of light is approaching the observer, light waves are shortened due to the motion of the source, and all colors are shifted toward the blue end of the spectrum. When the source is receding, light waves are lengthened and all colors are shifted toward the red end.

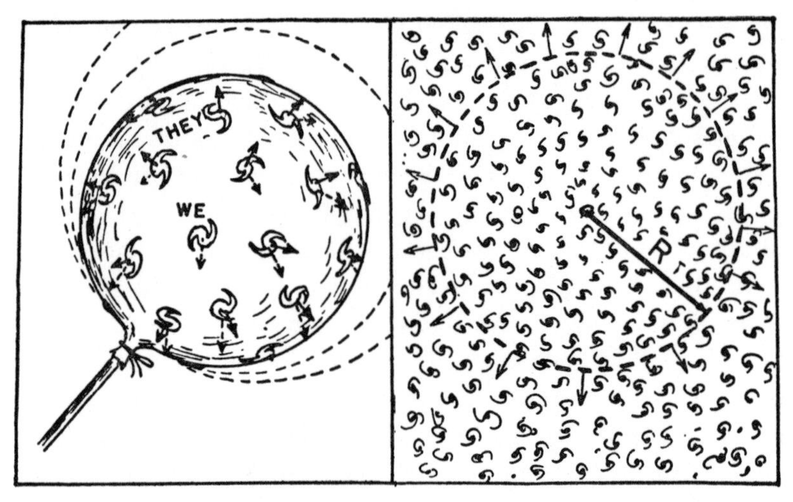

Fig. 3. The idea of uniform expansion

that all the others run away from him and he may be inclined to believe (incorrectly) that he is at the center of expansion. A three-dimensional model of an expanding space is given by the "jungle gym" often found on children's playgrounds. We need only imagine that the jungle gym extends in all directions without any limit, and is built of telescoping tubes so that the distances between the children sitting at various intersections are gradually increasing.

In addition to regular recession velocities resulting from the uniform expansion of the entire system, galaxies also possess individual random motion similar to the thermal motion of gas molecules. Since the velocities of that random motion are comparable in magnitude to the recession velocities of neighboring galaxies, the two kinds of motion may sometimes counteract each other and thereby produce confusing results. The component of the random velocity of a neighboring galaxy directed toward us may happen to be larger than the corresponding recession velocity of that galaxy. In that case the galaxy is moving toward us, and will show a violet shift in the lines of its spectrum. This is actually observed in the case of the Great Spiral Nebula of Andromeda. However, at greater distances the ever-increasing recession velocities soon become too large to be reversed by the random "thermal" motion of individual galaxies, and the expansion of the whole system becomes quite evident.

In addition to the mutual recession and the individual random motion, galaxies also show various degrees of rotation around their axes. A very few galaxies apparently do not rotate at all and possess regular spherical form. Other galaxies rotate with various speeds, shown by various degrees of elongation of their elliptical bodies. Most of the galaxies, however, rotate so fast that some of the material flows out from their equatorial bulges, forming the characteristic pattern of spiral arms. It is interesting to note that the average kinetic energy of galactic rotation is the same as the average kinetic energy of their random translatory motion. This fact agrees with the general law of statistical mechanics, which also holds for the translatory and rotational motions of molecules in an ordinary gas.

The theory of the expanding universe

It was first noticed by the Belgian scientist Georges Lemaître that the observed expansion of the system of galaxies agrees with the cosmological conclusions of the general theory of relativity. It is true that Einstein's original model of the so-called "spherical universe" is mechanically static and does not provide either for expansion or contraction. However, the Russian mathematician A. Friedmann pointed out that the static nature of Einstein's universe was the result of an algebraic mistake (essentially a division by zero) made in the process of its derivation. Friedmann then went on to show that the correct treatment of Einstein's basic equations leads to a class of expanding and contracting universes. In particular, the "spherical universe" originally conceived by Einstein was shown to be dynamically unstable, and apt to start contracting or expanding at the slightest provocation. Within the scope of this book we cannot delve either into a detailed mathematical treatment of the fundamental tensor equations of the Einstein theory, or into their application to the problems of cosmology. Fortunately, however, many properties of the relativistic cosmological models can be described and understood on the basis of classical Newtonian theory with the use of very little mathematics.

Imagine the limitless space of the universe with a multitude of galaxies scattered through it more or less uniformly. Can these galaxies stay at rest with respect to one another? Apparently not, since the force of Newtonian gravity acting between individual galaxies will pull them together so that the entire system will collapse (Fig. 4A). If we assume that the galaxies were not originally in a state of rest but were moving away from one another at high initial velocities, then the distances between the neighboring galaxies will continue to increase and the entire system will expand (Fig. 4B). We have here a situation comparable to that of a sports reporter inspecting a photograph showing a football high in the air above the heads of the players. Even though the picture does not show any motion, he is sure that the ball is either still rising or already falling. It could not have been hanging motionless above the field—unless, of course, it was resting on top of a pole or suspended on a string. In the same way the assumption of a static system of galaxies under mutual gravitational attraction necessitates the introduction of additional forces which prevent the galaxies from drawing together. We must imagine a system of struts (Fig. 4C) which counteract the forces of gravity and keep the galaxies apart. The mathematical apparatus of the general theory of relativity (which is nothing but a glorified generalization of the old Newtonian theory of gravity) leads to exactly that conclusion. In order to obtain a model of a static universe (this was before Hubble's discovery), Einstein was forced to introduce into his general equations an additional term containing

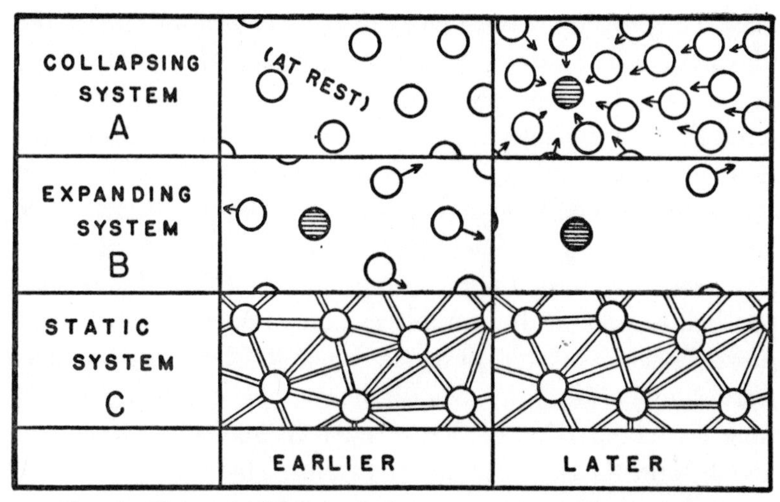

Fig. 4. Three possible states of a system of gravitating bodies

the so-called "cosmological constant" which was physically equivalent to the assumption of a *universal repulsion* between material bodies. In contrast to other physical forces which always decrease with distance, cosmological repulsion was assumed to be very weak over short distances but important at intergalactic distances.[3] Einstein's introduction of this force was not quite arbitrary, since, as it was shown later, this term represents a logical mathematical generalization from the original equations of general relativity. When the fact that our universe is not static but is rapidly expanding was recognized, the introduction of cosmological constant became superfluous. However, as we see later, this constant may still be of some help in cosmology, even though the primary reason for its introduction has vanished.

The discovery that our universe is expanding provided a

[3] The mathematical expression for this repulsive force acting on a particle of the mass m can be written as

$$F = -\tfrac{1}{3}\, c^2 \Lambda m r$$

where c is the velocity of light, Λ the new cosmological constant, and r the distance to the particle.

master key to the treasure chest of cosmological riddles. If the universe is now expanding, it must have been once upon a time in a state of high compression. The matter which is now scattered through the vast empty space of the universe in tiny portions which are individual stars must at that time have been squeezed into a uniform mass of very high density. It must have been subjected to extremely high temperatures, since all material bodies are heated when compressed and cooled when expanded. At present the possible maximum density of this compressed primordial state of matter is not accurately known. The nearest guess is that the over-all density of the universe at that time was comparable to that of nuclear fluid, tiny droplets of which form the nuclei of various atoms. This would make the original pre-expansion density of the universe a hundred thousand billion times greater than the density of water; each cubic centimeter of space contained at that time a hundred million tons of matter! In such a highly compressed state all the matter which is now within the reach of the 200-inch telescope must have occupied a sphere only thirty times as large as the sun. But since the universe is, and always was, infinite, the space outside of that sphere was also occupied by matter, the matter which now lies beyond the reach of the 200-inch telescope.

The fact that material occupying an infinite space can be squeezed or expanded and still occupy the same infinite space is one of the so-called "paradoxes of infinity." It is best illustrated by an example given by a famous German mathematician, David Hilbert, in one of his lectures.

"Imagine," said Hilbert, "a hotel with a finite number of rooms, all rooms being occupied. When a new client arrives, the room clerk must turn him down with regrets. But let us imagine a hotel with an infinite number of rooms. Even if all these rooms are occupied, the room clerk will be glad to accommodate a new customer. All he has to do is to move the occupant of the first room into the second, the occupant of the second into the third, the occupant of the third into the fourth, *und so weiter*. . . . Thus the new customer can get into the first room. Imagine now a hotel with an infinite number of rooms, all occupied," continued Hilbert, "and an infinite number of new customers. The room clerk will be glad to oblige. He will move the occupant of the first room into the second, the occupant of the second into the fourth, the occupant of the third into the sixth, *und so weiter*. . . . Thus every second room (all odd numbers) will now be free to accommodate the infinity of new customers."

In exactly the same way that an infinite hotel can accommodate an infinite number of customers without being overcrowded, an infinite space can hold any amount of matter and, whether this matter is packed far tighter than herrings in a barrel or spread as thin as butter on a wartime sandwich, there will always be enough space for it.

What started the expansion?

We can now ask ourselves two important questions: why was our universe in such a highly compressed state, and why did it start expanding? The simplest, and mathematically most consistent, way of answering these questions would be to say that *the Big Squeeze which took place in the early history of our universe was the result of a collapse which took place at a still earlier era, and that the present expansion is simply an "elastic" rebound which started as soon as the maximum permissible squeezing density was reached.* As was indicated in the previous section, we do not know exactly what was the density reached at the maximum of compression, but according to all indications this density could have been very high indeed. Most likely the masses of the universe were squeezed together to such an extent that any structural features which may have been existing during the "pre-collapse era" were completely obliterated, and even the atoms and their nuclei were broken up into the elementary particles (protons, neutrons, and electrons) from which they are built. Thus nothing can be said about the pre-squeeze era of the universe, the era which may properly be called "St. Augustine's era," since it was St. Augustine of Hippo who first raised the question as to "what God was doing before He made heaven and earth." As soon as the density of the masses of the universe reached its maximum value, the direction of motion was reversed and the expansion started, so that very high densities could have existed only for a very short time. During the earlier or later stages of this expansion, various differentiation processes must have taken place in the cosmic masses, processes which resulted in the present highly complex structure of our universe.

The date of the Big Squeeze

We have seen that Hubble's observations indicate that the galaxies scattered through the vast space of the universe seem to be running away from us, and the farther they are the faster they run. The relationship between recession velocity and distance is given by Hubble's law (see Appendix, pages 140–42):

$$[\text{recession velocity}] = \text{constant} \cdot [\text{distance}]$$

In his original studies Hubble found that the numerical value of the constant is $1.8 \cdot 10^{-4}$, if the distances are expressed in light-years and the velocities in kilometers per second. If, as is customary in physics, we measure distances in centimeters and velocities in centimeters per second the original Hubble's value of the constant becomes $1.9 \cdot 10^{-17}$. In order to find at what date in the past all galaxies were packed tightly together, we need only to divide their present mutual distances by the velocities of their recession. Since the recession velocities are proportional to the distances, the result of the division is always the same, no matter whether we take two neighboring or two distant galaxies. It is simply the inverse value of the constant in Hubble's law. Thus for this date Hubble obtained the value:

$$\frac{1}{1.9 \cdot 10^{-17}} = 5.3 \cdot 10^{16} \text{ sec} = 1.7 \cdot 10^{9} \text{ yr.}$$

Comparing this figure with the various other estimates of the age of the universe, we find that it falls uncomfortably short of the average. In particular it is only about one-half of the figure obtained by Holmes from the study of the relative abundances of radiogenic lead isotopes in the rocks forming the crust of the earth. How could the universe, which is less than two billion years old, contain the rocks which are over three billion?

This discrepancy pestered proponents of the theory of an expanding universe for several decades, from the original work of Hubble until the early nineteen-fifties. One possibility suggested by Lemaître was to introduce the cosmological constant originally used by Einstein in his attempt to construct a model of a static universe. This cosmological constant corresponds physically to a repulsive force acting between the galaxies over long distances, and increasing in direct proportion to the distance. The presence of such a force would make the universe expand with ever-increasing velocity and shift the position of the zero point in time. Indeed, if the expansion process is accelerated, the recession velocities of the neighboring galaxies would have been smaller in the past than today, so that the date of the beginning would be shifted back in time. Assuming such a small numerical value as 10^{-33} sec^{-1} for the cosmological constant Λ, one could bring Hubble's original value into agreement with the geological estimate.

Another much more radical modification of the expansion theory was proposed by three British mathematical astronomers, H. Bondi, T. Gold, and F. Hoyle. They based their theory on the principle of four-dimensional isotropy of the universe, according to which the behavior of the universe in time must be the same as its behavior in space. Since, according to astronomical observations, the universe looks the same no matter how far we go in any direction in space, this principle demands that the universe must look the same no matter how far we go forward or backward in time. In order to reconcile this conclusion with the observationally established fact of the dispersion of galaxies, B., G., & H. had to postulate that the continuous thinning of matter in the space of the universe, caused by continuous expansion, is compensated by continuous creation of new matter taking place uniformly throughout intergalactic space. This postulate does not, as was sometimes stated, contradict the law of conservation of matter, since the amount of matter within a given volume always remains constant. We simply have here one of the paradoxes of infinity, which falls into the category of the hotel example by David Hilbert. To compensate fully for the expansion no more is required than the creation of one new hydrogen atom per liter of space once every billion years, so that a creative genius would not overstrain himself doing the job. According to these views the older galaxies are gradually receding farther and farther, but all the time new galaxies are being formed by condensation of newly created matter in the widening spaces between the old ones. Thus the show goes on, without beginning and without end.

If we were to make a motion picture representing the views of B., G., & H., and run it backward, it would seem at first that all the galaxies on the screen were going to pile up as soon as we reached the date 1.7 billion years ago. But, as the film continues to run backward, we would notice that the nearby galaxies, which were approaching our Milky Way system from all sides, threatening to squeeze it into a pulp, would fade out into thin space long before they became a real danger. And, before the second-nearest number could converge on us, our own galaxy would fade too. While this point of view provides for the origin and evolution of individual galaxies, it considers the universe itself as being eternal, though with a constantly changing galactic population. While the conventional picture of an expanding universe can be compared with a group of people all born in the same year and aging together, the B., G., & H. view corresponds to a community in which children are born while old people die, so that the composition of the group remains invariant. In such a case all ages are present and there is no contradiction between Hubble's value of 1.7 billion years and the estimated age of our own galaxy.

In the early nineteen-fifties, however, it developed that neither of the two previously described possibilities is required to remove the discrepancy between the value of Hubble's constant and the other estimates of the age of the universe. Mount Wilson astronomer W. Baade, revising previous work on the intergalactic distances, has found an error in the method used in their measurement. These distances are measured by observing pulsating stars, or Cepheids, located in the galaxies which are not too far from us for individual bright stars to be visible. It was established by observation of the Cepheids within our galaxy that the period of their pulsations depends on their absolute brightness in a well-defined way. Thus, if one sees a Cepheid in some galaxy and measures the period of its light changes, one can tell what its absolute brightness is. Comparing that absolute brightness with the observed visual brightness, one can find out how far away this star and the entire galaxy are. In his studies of the stars of the Milky Way, Baade found that there are two rather different types of stellar population: one belonging to the spiral arms (Type I) and one to the central body of the galaxy (Type II).[4] It turned out that the Cepheids belonging to these two different types of populations are also different and possess quite different luminosities corresponding to the same pulsation period. The mistake of the earlier distance

[4] See Chapter IV.

measurements lay in the fact that astronomers, not being aware that there are two different types of Cepheids, were applying the brightness-period relation established for the Cepheids of one type to the Cepheids of other types. Putting things straight, Baade found that the earlier estimates of the distances of various galaxies were too small by a factor of about 2.5. Thus, with a flick of his wrist, Baade increased the range of the Palomar Mountain 200-inch telescope from the previously estimated value of one billion light-years to two and a half billion light-years! What is more important, the time necessary for the galaxies to recede to their present distances went up from $1.7 \cdot 10^9$ years to about $4.3 \cdot 10^9$ years. The disagreement between the age of the universe obtained on the basis of galactic recession and that obtained by all other methods exists no more.

This progress killed the theory of the accelerated expansion of the universe; if this were the case the age of the universe estimated on the basis of galactic recession would be much greater than the values obtained by other methods. It also made unnecessary the B., G., & H. hypothesis of continuous creation of matter, but it by no means eliminated it as a possibly correct theory. Indeed, since according to this hypothesis the universe may contain the galaxies of all ages, one can always find a galaxy of a proper age to call it our Milky Way. The test of the B., G., & H. theory rests now on comparing the ages of the galaxies populating the space around us. According to that theory they must show a large variety of ages, whereas the conventional theory would expect them to be equally old. Furthermore, according to the B., G., & H. theory, the average age of the galaxies must be equal to one-third of the inverse of Hubble's constant, that is, $1.4 \cdot 10^9$ years, which is much less than the age of our Milky Way. If this is true, the position of our Milky Way among the neighboring galaxies must be that of a middle-aged man sitting in a group of schoolboys. One can expect the galaxies to show age in somewhat the same way that man does. In the older galaxies the brightest stars must have exhausted their nuclear energy supply and faded out; so that the older the galaxy, the fainter are the brightest stars it contains. If it were true that the Milky Way is much older than all the neighboring galaxies, the stars in our neighbors would shine much more brightly than the stars of the Milky Way. Astronomical observations show that this is not the case, which suggests that the theory of B., G., & H. may not correspond to reality.

Will the expansion ever stop?

The question of whether the present expansion of our universe will continue forever or will at some time in the future stop, and then turn into a collapse, resembles the question of whether a rocket fired from the surface of the earth will continue to travel in space or stop and fall back on our heads. In the case of the rocket everything depends on the velocity which it acquired. If its velocity is more than 11.2 kilometers per second (the so-called "escape velocity,") the rocket will defy gravity and never come back; for any lesser velocity the rocket will inevitably come to a stop at a certain altitude and then fall back. In the first case we say that the kinetic energy of the rocket was greater than the potential of the terrestrial gravitational field; in the second case the situation is reversed.

The case of galaxies receding from one another with given velocities against the forces of mutual gravity which try to pull them together is similar to the example of the rocket. The question is simply whether the inertial force of the galactic recession or the pull of their gravitational fields is more powerful. Simple calculations, presented in the Appendix, indicate that at the present epoch gravitational pull between galaxies is negligibly small as compared with their inertial velocities of recession. Here we have a case similar to that of a rocket moving away from the earth with a velocity far higher than escape velocity. The distances between the neighboring galaxies are bound to increase beyond any limit, and *there is no chance that the present expansion will ever stop or turn into a collapse.*

Is our universe finite or infinite?

This seems to be the proper moment to raise the question of the total size of our universe. Is it finite, having a volume of so many cubic feet, or so many cubic light-years, as Einstein once suggested, or does it extend without limit in all directions as visualized by good old Euclidian geometry? Within the distance of 1 billion light-years covered by the 200-inch telescope of Palomar Observatory the galaxies (about 1 billion of them) seem to be scattered through space in a more or less uniform fashion. But suppose astronomers go on building 400-inch, 800-inch, 1600-inch, etc., telescopes. What will they find? Einstein's general theory of relativity and gravitation leads to two possible mathematical alternatives.

The first is that the space of the universe may "curve in" in the manner of the surface of the earth (positive curvature) and finally close upon itself in an "antipodal point" (Fig. 5, top left). This is Einstein's closed universe (which can be either static or expanding), in which, barring space obscuration, one would be able to inspect the haircut on the back of one's neck, if one could manage to wait for several billion years while the light travels around the universe. Or else the space of the universe may "curve out," like the surface of a western saddle (Fig. 5, lower left).

Which of these two possibilities must be ascribed to our universe, only observational evidence can decide. And the existing evidence seems to be strongly in favor of the second possibility, namely, the limitless infinite universe. From the purely mathematical point of view, whether the geometry of the universe is "open" or "closed" will directly determine its behavior in the course of time. It can be shown that a closed Einsteinian universe can expand only to a certain limit, beyond which the ex-

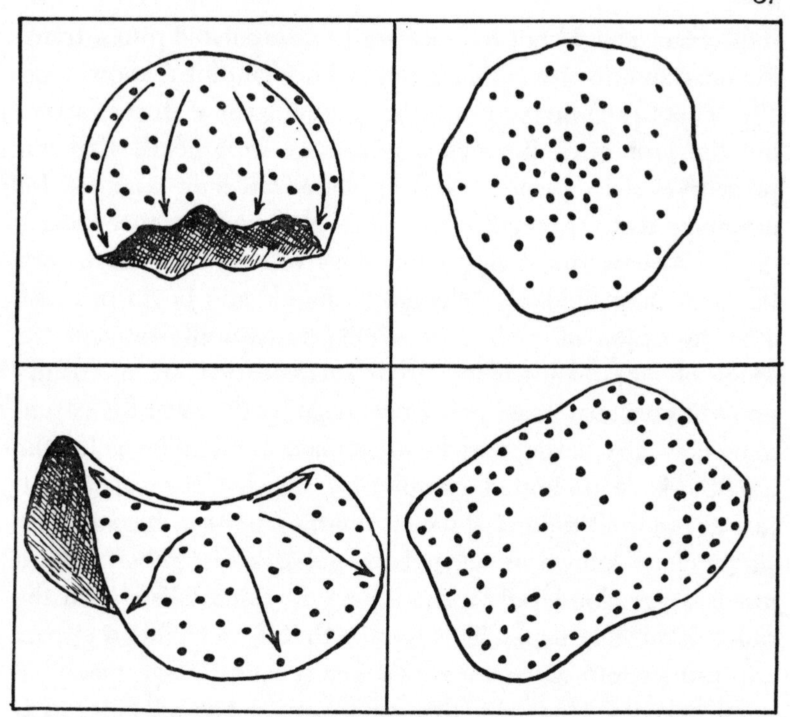

Fig. 5. Surfaces of positive and negative curvature (spherical and saddle surfaces), and what happens to each when we try to flatten it on a table

pansion will go over into contraction. But an open universe is bound to expand forever without any limit. One can say that if the universe is closed and periodic in space (as the surface of a sphere), it is also periodic in time and is subject to alternating expansions and contractions (pulsating universe). On the other hand, if the universe is open and aperiodic in space (as the surface of a paraboloid), it will not repeat itself in time either. Thus the same arguments which, in the previous section, led us to accept the *limitless expansion* of the universe also lead us to accept its *limitless extension.*

Of course these conclusions, which are based entirely on observational evidence within a sphere with a radius of 2.5 billion light-years, should not be necessarily extrapolated into a trans-Palomarian universe which is beyond our empirical knowledge. We cannot exclude the possibility that at some distance farther out the properties of space may change so radically that our present knowledge no longer applies. One may imagine, for example, that, although within the observed distances space does not show the slightest tendency to close, at greater distances it may suddenly "change its mind" and begin to close. The discussion of such a possibility is naturally outside the scope of empirical science. In this connection an interesting point of view expressed many years ago by the Swedish astronomer C. V. L. Charlier may be mentioned. It might be called the hypothesis of unlimited complexity. Charlier suggested that, just as the multitude of stars surrounding our sun belongs to a single cloud known as our galaxy, galaxies themselves form a much larger cloud, only a small part of which falls within the range of our telescopes. This implies that if we could go farther and farther into space we would finally encounter a space beyond galaxies. However, this supergiant galaxy of galaxies is not the only one in the universe, and much, much farther in space other similar systems can be found. In their turn these galaxies of galaxies cluster in still larger units *ad infinitum!* Intriguing as it is, this picture of an ever-increasing aggregation of matter is unfortunately outside the possibility of observational study.

Nebular counts, and the confusion between distances and ages

There is another method for finding out whether the space of the universe curves "in" or "out," a method based on pure geometry and quite independent of the theory of expansion in time. It is connected with the question of how much space is available within a given distance and can be best understood by referring to a two-dimensional example. Suppose we cut out a circular piece of leather from an ordinary football and try to flatten it on the surface of a table (Fig. 5, top right). It is obvious that this cannot be done without stretching the leather at the outer parts of our piece. Since the original surface curved in, there is not enough material at its peripheral portions. If we now repeat the same experiment with a piece of leather cut from a Western saddle (Fig. 5, lower right), the situation is entirely different. There is too much leather at the rim and we would have to shrink it to make it flat. If the two surfaces are covered with an originally uniform pattern of dots, we find that after flattening there will be a rarification of dots near the rim of the spherical surface and a crowding of dots near the rim of the saddle surface. Speaking mathematically, the area of a circle drawn on a spherical surface increases more slowly than the square of the radius; on a saddle surface it increases faster than the square of the radius.

A similar situation exists in the case of three-dimensional curved space, even though, being inside of that space, we cannot visualize it as easily as we can two-dimensional surfaces which we can look at from the outside. Instead of dots on the leather we now have galaxies in space, presumably in uniform distribution. If space is finite and curves in, the number of galaxies must increase more slowly than the cube of that distance. If space is infinite and curves out, the number of galaxies should increase faster than the cube of that distance.

Such "galactic counts" have been made by Hubble, who found, in contradiction to the conclusions we reached in the previous section, that the space of the universe is curving in, and very fast indeed. However, Hubble's result depends entirely on the correctness of the estimates of galactic distances. Since according to the results of W. Baade mentioned above, these distances are actually two and a half times as large as previously thought, galactic counts lead to the opposite result, and the universe is found to be open and infinite. It must also be remembered that all the estimates of galactic distances are based on the assumption that the galaxies possess a constant luminosity; in fact, the distances are measured simply by using the inverse-square law for the visual brightness of a distant source. Since we see distant galaxies as they were hundreds of millions of years ago, the results of galactic counts will be substantially different if galaxies change their luminosities with time. As the stellar population of galaxies may be expected to evolve with time (see Chapter V), it is logical for us to assume that the luminosities of galaxies in the past differed from those of today.

We cannot use Hubble's method of galactic counts for the study of the curvature of space unless we first know how much and how fast the galaxies change their luminosity with age. On the other hand, a fruitful study of evolutionary changes of galactic luminosities is possible only if we have a reliable method for estimating their distances, which in turn requires a knowledge of the geometry of the universe. The only way to proceed seems to be to estimate the geometry of the universe from the theory of expansion (as was done in the previous section) and to use these results to decipher the observations pertaining to evolutionary changes of galactic luminosities.

Early stages of expansion

Mathematical studies of the expansion process, presented in some detail in the Appendix, indicate that the constant in Hubble's law changes slowly with the progressing evolution of the universe. For comparatively early stages of the expansion, the value of Hubble's constant is connected with the mean density of the universe by the relation:

$$[\text{Hubble's constant}]^2 = 5.8 \cdot 10^{-7}\ [\text{mean density}]$$

It must be pointed out that the mean density of the universe includes not only the density of ordinary matter but also the mass-density of radiation (light visible and invisible) filling space. We know that, according to Einstein's famous principle of "equivalence of mass and energy," radiation possesses a certain weight which can be expressed numerically by dividing its energy by the square of the velocity of light. In everyday life the weight of radiation is so small that it might as well be neglected. Specifically, the weight of the visible light in a brightly illuminated room is negligible when compared with the weight of the air filling the same room. In the cosmos, however, the situation is different, not so much because the weight of radiation is higher, as because the mean density of matter is so low. According to a well-known formula of classical physics (the so-called Stephan-Boltzmann formula), the amount of radiant energy per unit volume of space at temperature T (absolute temperature counted from the zero point at —273 degrees centigrade) is equal to $7.6 \cdot 10^{-15} T^4$. Dividing this by the square of the velocity of light ($c^2 = 9 \cdot 10^{20}$) we find for the weight of that radiant energy the value of $8.5 \cdot 10^{-36} T^4$ grams per cubic centimeter. At normal room temperature (about 300 degrees absolute), the weight of radiation (in this case heat rays) is only 10^{-25} grams per cubic centimeter. In interstellar space, which is heated by stars and has a constant temperature of about 100 degrees absolute (near the temperature of liquid air), the density of radiation (very cold heat rays!) is 10^{-27} grams per cubic centimeter. Small as this is, it forms about 0.1 per cent of the density of interstellar gas, which is 10^{-24} grams per cubic centimeter.

From the laws of classical physics, we can derive the fact that the density of radiation in an expanding volume will drop faster than the density of matter in the same volume.[5] We then have to assume that *during the earlier stages of expansion the*

[5] If the edge of a cubical container is increased by a factor a, its volume will increase by a factor a^3, and the density of matter in it will decrease

weight of the radiation in each volume of space exceeded that of the matter in the same volume. During these epochs ordinary matter did not count, and the main role was played by intensely hot radiation.

One may almost quote the Biblical statement: "In the beginning there was light," and plenty of it! But, of course, this "light" was composed mostly of high-energy X rays and gamma rays. Atoms of ordinary matter were definitely in the minority and were thrown to and fro at will by powerful streams of light quanta.

The relation previously stated between the value of Hubble's constant and the mean density of the universe permits us to derive a simple expression giving us the temperature during the early stages of expansion as the function of the time counted from the moment of maximum compression. Expressing that time in seconds and the temperature in degrees (see Appendix, pages 142–43), we have:

$$\text{temperature} = \frac{1.5 \cdot 10^{10}}{[\text{time}]^{1/2}}$$

Thus when the universe was 1 second old, 1 year old, and 1 million years old, its temperatures were 15 billion, 3 million, and 3 thousand degrees absolute, respectively. Inserting the present age of the universe ($t = 10^{17}$sec) into that formula, we find

$$T_{\text{present}} = 50 \text{ degrees absolute}$$

which is in reasonable agreement with the actual temperature of interstellar space. Yes, our universe took some time to cool from the blistering heat of its early days to the freezing cold of today!

While the theory provides an exact expression for the tem-

---

by the factor a^3. But the temperature of radiant energy in that volume will decrease by the factor a (Wien law), so that its density drops by a factor a^4 (according to the Stephan-Boltzmann law).

perature in the expanding universe, it leads only to an expression with an unknown factor for the density of matter. In fact, one can prove (see Appendix) that

$$[\text{density of matter}] = \frac{\text{constant}}{[\text{time}]^{3/2}}$$

We see in Chapter III that the value of that constant may be obtained from the theory of the origin of atomic species.

CHAPTER III

The Making of Atoms

Natural abundance of atomic species

As has been pointed out several times, it seems reasonable to assume that the various atomic species were formed during the very early period of expansion when all the matter in the universe was still uniformly squeezed to extremely high density and subjected to very high temperatures, providing favorable conditions for all kinds of nuclear transformations. Since the relative amounts of different atomic species produced during this "universal cooking era" must have been determined by physical conditions then prevailing, our knowledge of the relative abundance of chemical elements and their isotopes in nature should enable us to reconstruct the picture of the early stages of expansion. Thus the table of relative abundances of atomic species may be thought of as the oldest document pertaining to the history of our universe.

Relative abundances of elements have been exhaustively studied by a great many geophysicists and astrophysicists, and today we have a large amount of material on that subject. The bulk of the data comes from chemical analysis of the earth's crust and of the meteorites which presumably represent the fragments of a former planet that once moved between the orbits of Mars and Jupiter. These data were supplemented by spectral analysis of the sun, the stars, and of the diffuse material scattered through interstellar space. The most important result of these studies is the fact that *the chemical constitution of the universe is surprisingly uniform.* It was found that about 55 per cent of cosmic matter is hydrogen and about 44 per cent helium; the remaining 1 per cent accounts for all heavier elements, in the same proportions as we find them on earth. Considering the general figures for cosmic abundance, our earth represents a remarkable exception, almost lacking in hydrogen and helium, which are the main constituents of matter in the universe.[1]

The scarcity of hydrogen, helium, and other rare gases on our planet is purely a local effect, the result of the circumstances under which the birth of the earth took place. As Chapter IV shows, the formation of our planetary system started by a process of aggregation of interplanetary dust, resulting in what may be called primeval planetary cores. This dust, floating in the interstellar gas of mixed hydrogen and helium, has about the same constitution as the dust clouds raised by a bulldozer working on a mountain road, and its aggregation produced the rocky bodies of our earth, of Mars, Venus, and other minor planets. Gaseous material filling interstellar space did not participate in this process until the original rocky bodies had grown large and massive enough to capture these gases by gravitational force. In order to be able to capture interstellar gas, a planetary core had to grow to be several times as heavy as our earth. Neither the earth, nor Mars, nor Venus ever grew large enough for that (presumably because the supply of dust in their neighborhood was ex-

[1] Everybody knows that helium is a very scarce element on earth, but it would seem at first glance that hydrogen is quite abundant. This impression is, however, entirely due to the fact that we happen to live in the midst of the main hydrogen deposit, namely, the moisture of the atmosphere, and near the oceans which cover two-thirds of the earth's surface. If we remember that the depth of oceans is negligibly small as compared with the diameter of our globe, and that the granitic and basaltic rocks which form the main body of the earth do not contain any hydrogen, we will realize that hydrogen is really quite rare on our planet.

hausted) and thus they remained rocky planets as we know them today. The primeval cores of Jupiter, Saturn, and other major planets, however, exceeded the critical mass limit and surrounded themselves with heavy atmospheres of hydrogen and helium. Thus, according to the studies of Harrison Brown, the rocky core of Jupiter, being similar to our earth but about six times as heavy, amounts to only about 2 per cent of the total mass (300 earth-masses) of the planet. This central core is covered by layers of frozen water, ammonia, and methane, which together account for another 8 per cent. The remaining 90 per cent of the giant body of Jupiter consists of a highly compressed mixture of hydrogen and helium which almost reaches the density of water near the hidden surface of the central core. This internal structure of Jupiter, superimposed on its regular photograph, is shown in Plate III. Only small planets and satellites, then, are exempt from the rule that the universe is composed of 55 per cent hydrogen, 44 per cent helium, and 1 per cent "other atoms."

Figure 6 represents the cosmic abundances of atomic species by a diagram based on the classical work of the Norwegian geochemist, V. M. Goldschmidt, supplemented by more recent data obtained by Harrison Brown from the analysis of meteorites. We see clearly that cosmic abundances drop very rapidly with increasing atomic weight. Such elements as silver or molybdenum, located halfway up the periodic system, are present in the amount of only about one part in several billions. The striking feature of the natural abundance curve is that *after reaching the elements with an atomic weight of about 100, the curve levels out, indicating nearly equal abundances for all elements in the upper half of the periodic system.* This surprising shape of the empirical curve, its original rapid (exponential) fall, followed by horizontal continuation, obviously contains an important hint about the conditions under which the atoms originated. Any theory which claims to give a consistent picture of the nu-

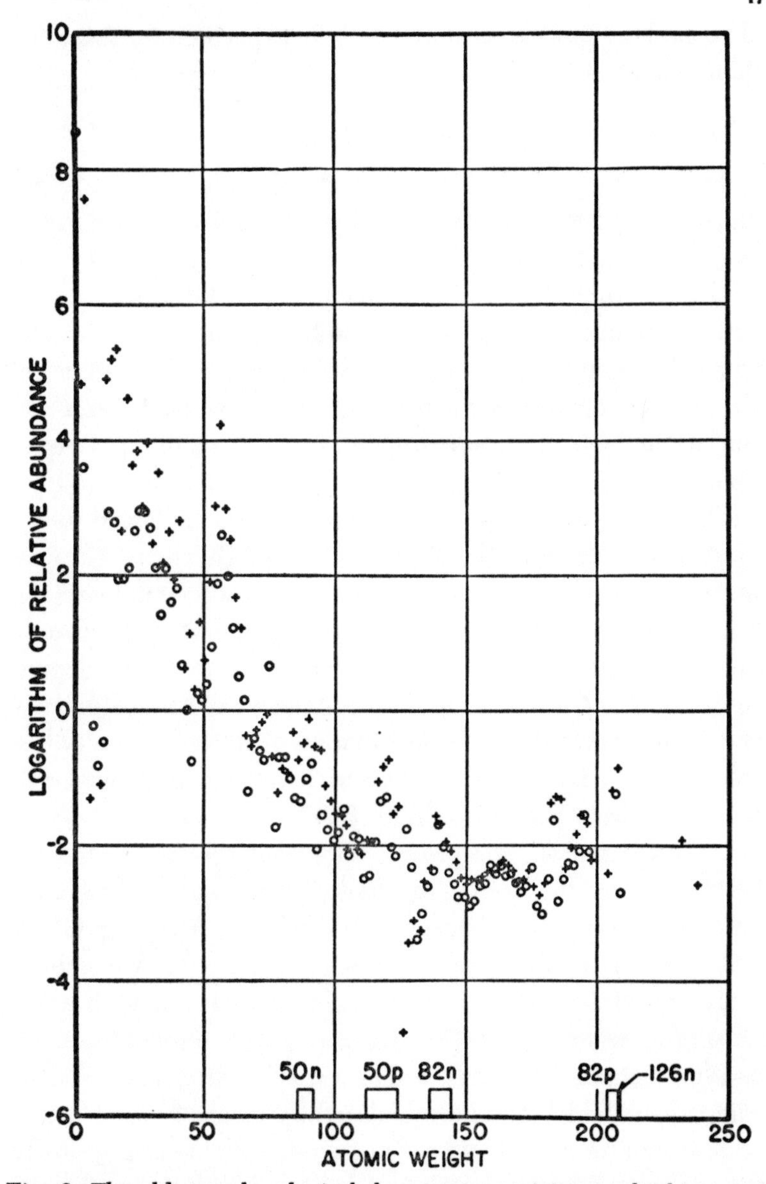

Fig. 6. The oldest archaeological document pertaining to the history of the universe. Observed abundances of various atomic species in the universe, plotted (in logarithmic scale) against their atomic weights. The boxes indicate "magic numbers"

clear cooking process must also be able to account for this abundance curve.

Hypothesis of frozen equilibrium

Earlier interpretations of the empirical curve of natural abundances were based on the simple—and in a way a most logical—assumption that the observed distribution represents some kind of chemical (or rather alchemical) equilibrium between various atomic species. This equilibrium must have existed when the temperature of matter was sufficiently high to permit all kinds of thermonuclear reactions and must have been "frozen" when the temperature dropped as the result of rapid expansion.

The notion of "frozen equilibrium" can be explained with a simple example from the field of physical chemistry. Suppose we have a certain amount of water in a closed container (Fig. 7 A). If we heat it to a sufficiently high temperature (above the so-called "critical temperature" which for water is 374 degrees centigrade), all the water in the container will turn into vapor; at this temperature the thermal motion of water molecules is so fast that cohesive forces between them can no longer hold them together (Fig. 7 B). With further increases of temperature the thermal motion grows violent enough so that the collisions between the rushing molecules begin to break them apart into separate atoms of hydrogen and of oxygen (Fig. 7 C). At a temperature of a few thousand degrees our retort will contain a mixture of free atoms of hydrogen and oxygen with a few remaining water molecules. For any given temperature and pressure (or density) there will be a specific proportion of these components which can be calculated by means of the classical equilibrium formula. If we now cool our container very slowly, the process will reverse itself and we'll end with the water with which we started. If, however, we cool the container very rapidly (by pouring liquid air over it or by expanding the gas to a much larger volume), the recombination process must take place in a hurry. Some atoms will still have time to find the proper mates, and some H_2O molecules as well as H_2 and O_2

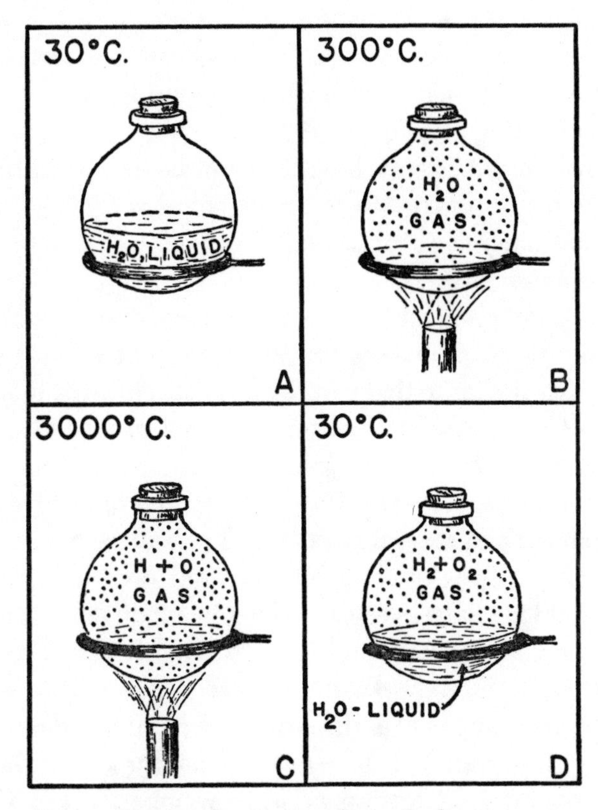

Fig. 7. How to obtain an explosive hydrogen-oxygen mixture by means of frozen equilibrium

molecules will be formed (Fig. 7 D). But for normal temperature, there should be only water. Hence the mixture of water plus free oxygen and hydrogen represents the equilibrium for a much higher temperature. Such a mixture is in "metastable equilibrium": if we put a match to it, an explosion will follow and the gas will turn back into water vapor, later condensing to water.

This example presents a good analogy of what could have happened during the rapid expansion of the originally highly heated matter in the universe. This might be the explanation of the fact that the present assortment of atomic species contains many nuclei which can still react with one another, thus liberating vast amounts of hidden nuclear energy. In fact, if the assembly of existing atomic species were now in a state of equilibrium, any Atomic Energy Project would be as impossible as a project to obtain energy from the chemical reactions between the various minerals (coal and oil, of course, excluded) forming the crust of our earth.

Equilibrium proportions of various atomic species for any given conditions of temperature and density can be calculated by a formula similar to that used by physical chemists in the calculation of the states of equilibria between molecules. All one has to do is to substitute nuclear binding energies (amounting to several million electron volts) for molecular binding energies (which amount to several electron volts) and of course to use much higher values for temperatures and pressures. Calculations of that kind have been carried out by many scientists, notably by S. Chandrasekhar and V. L. R. Henrich, and more recently by O. Klein, G. Beskow, and L. Treffenberg. It was found that the descending part of the curve (excluding all minor irregularities of the empirical curve) can be interpreted as the state of frozen equilibrium, corresponding to a temperature of 8 billion degrees and a density ten million times that of water. However, these calculations ran aground in their attempt to follow their initial success into the region of the heavier elements. Instead of leveling out, as the empirical abundance curve does, the theoretical curve continued its steep descent (Fig. 8). The empirical abundances of heavy elements are billions of billions of times greater than the values predicted by the theoretical

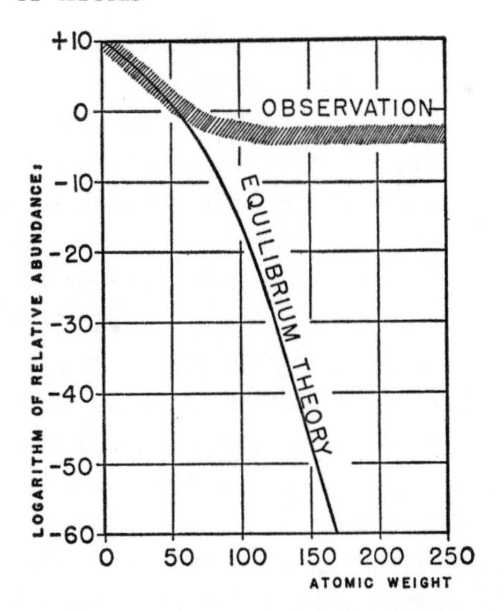

Fig. 8. Shortcomings of the theory of frozen equilibrium

curve. This negative result was not due to any specific assumption made in the calculations but followed directly from the nature of the equilibrium hypothesis.[2]

This mishap of the equilibrium theory led to various attempts to patch it up. Thus, for example, Chandrasekhar and Henrich themselves suggested that light and heavy elements were cooked at different times during the expansion process. According to that hypothesis, heavy elements were cooked and "frozen" at an earlier date when the temperature was very high, and the cooking of lighter elements continued for a longer time at lower temperatures. Such an assumption is, however, incon-

[2] In this theory, relative abundances are given crudely by the expression:

$$\lg(\text{ABUNDANCE}) = E_A/kT$$

where E_A is the total binding energy of nuclei of atomic weight A, k the so-called Boltzmann constant, and T the absolute temperature. Since throughout the periodic system the nuclear binding energies are roughly proportional to the atomic weights, the lg (abundance)-versus-AW curve must go straight down all the way from hydrogen to lead.

sistent with what nuclear physics knows about nuclear reaction rates under the conditions in question.

The only way out would be to assume that different atoms were cooked at different places and under different temperature and pressure conditions. Thus, Klein, Beskow, and Treffenberg in Sweden assume that different chemical elements were formed at different depths in some primordial (no longer existing) stars, which later exploded, spreading the material all over space. A similar opinion is expressed by the Dutch astronomer G. B. van Albada and the British astronomer Fred Hoyle, who prefer the job to be done in the now existing stars: van Albada in the so-called Red Giants, and Hoyle in the exploding supernovae. Later writers were unable, however, to calculate the relative abundances to be expected on the basis of their theories, mostly because of the immense complexity of the picture involved. It must always be possible to reproduce *any* given abundance curve of elements by using suitably chosen conditions for explaining its different parts. What van Albada and Hoyle demand sounds like the request of an inexperienced housewife who wanted three electric ovens for cooking a dinner: one for the turkey, one for the potatoes, and one for the pie. Such an assumption of heterogeneous cooking conditions, adjusted to give the correct amounts of light, medium-weight, and heavy elements, would completely destroy the simple picture of atom-making by introducing a complicated array of specially designed "cooking facilities."

Hypothesis of the primeval atom

Another possible explanation of the origin of atomic species was proposed by the Belgian scientist Georges Lemaître who formulated the so-called "hypothesis of the primeval atom" twenty years ago. A more correct name would be "primeval nucleus," since Lemaître suggested that before the expansion started all the matter of the universe was in the state of dense

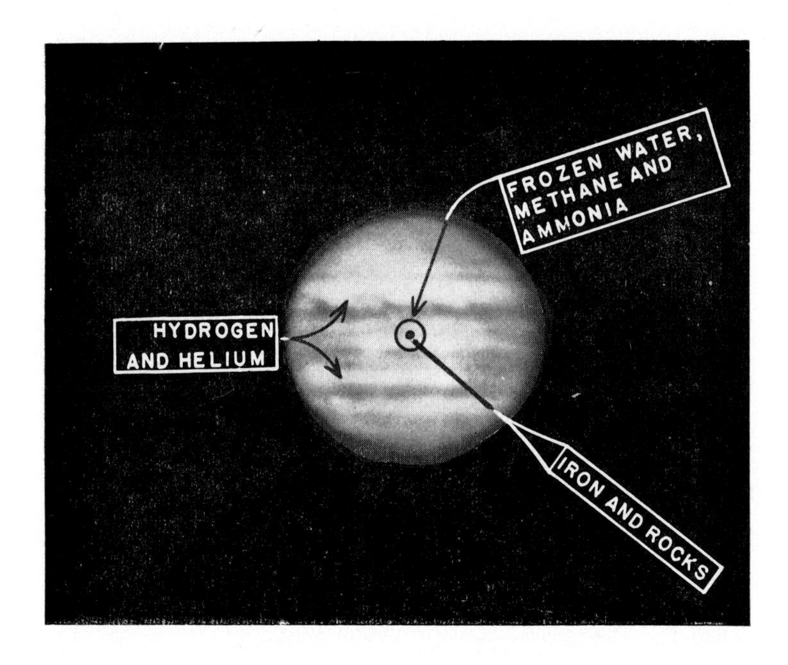

Plate III. Jupiter. The superimposed arrows indicate the internal structure of the planet

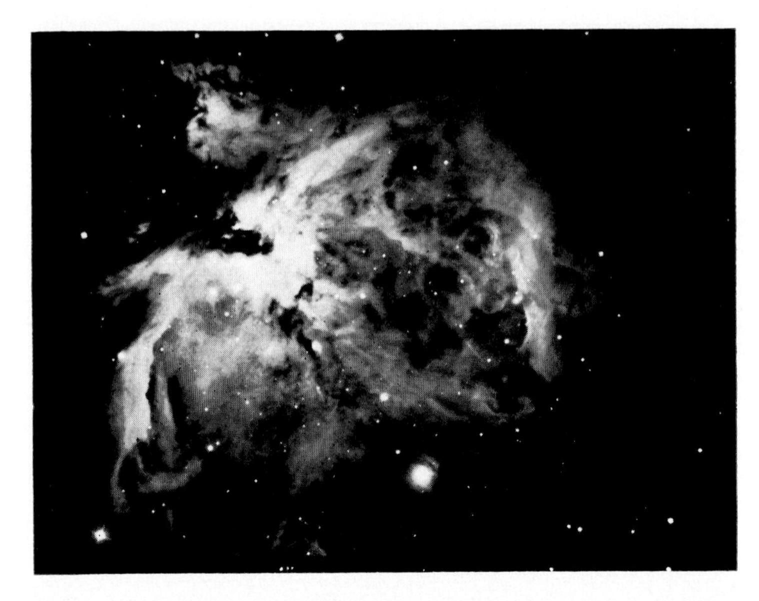

Plate IV. The Great Nebula in Orion, illuminated by nearby stars

Plate V. Dark nebula in Aquila, obscuring background stars

nuclear fluid, forming a giant nucleus similar to ordinary atomic nuclei but of course much larger. This assumption implies that at the beginning of expansion the temperature of matter was comparatively low (below the "critical temperature" of nuclear fluid), so that the thermal motion of nucleons was not strong enough to break the cohesive bonds binding them together into one continuous fluid. When expansion started, the original fluid became mechanically unstable and began to break up into fragments of all possible sizes. Here is the description of that process in Lemaître's own words:

> The atom world broke up into fragments, each fragment into still smaller pieces. Assuming, for the sake of simplicity, that this fragmentation occurred in equal pieces, we find that two hundred and sixty successive fragmentations were needed in order to reach the present pulverization of matter into our poor little atoms which are almost too small to be broken farther. The evolution of the world can be compared to a display of fireworks that has just ended: some few red wisps, ashes, and smoke. Standing on a cooled cinder, we see the slow fading of the suns, and we try to recall the vanished brilliance of the origin of the worlds.[3]

The author of these spectacular views did not choose to follow up the details of the fragmentation process by means of strict mathematical analysis. This chore fell on the shoulders of Maria Meyer and Edward Teller of Chicago, who arrived, quite independently, at similar views concerning the origin of atomic species. These writers do not discuss the very beginning of the break-up process in its relation to the general theory of the expanding universe. They begin with the stage at which the individual fragments were reduced to a size of several miles in diameter and to a mass comparable to that of an average star[4] (Lemaître's "atom stars"). At the moment such fragments were

[3] In *Revue des Questions Scientifiques,* November 1931.

[4] Since nuclear fluid has a density of 10^{14} times that of water, the radius of a nuclear droplet having the mass of the sun would be about 9 miles.

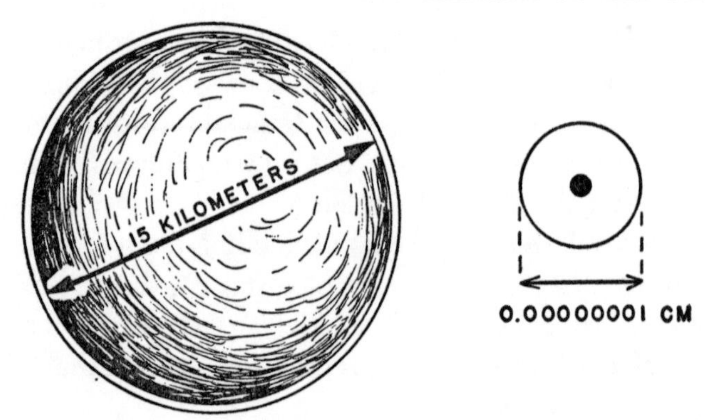

Fig. 9. A hypothetical superatom and an ordinary atom. The thickness of the atmosphere in the first and the size of the nucleus in the second are greatly exaggerated

formed by the mechanical break-up process of the originally neutral nuclear fluid, they must have consisted exclusively of neutrons. However, as the result of spontaneous [neutron $\rightarrow$ proton + electron] transformations, these original "polyneutrons" must soon have become positively charged and surrounded by thin electronic atmospheres. These primeval atoms were rather similar to the ordinary atoms of today, except for their giant size and the fact that the thickness of their electronic atmospheres (only 10^{-10} centimeters) was negligibly small (Fig. 9) as compared with the diameters of their nuclei.

Studying the mechanical stability of such superatoms, Meyer and Teller were able to show that the surface of the giant nucleus would be quickly covered by a multitude of miniature pimples or buds about 10^{-12} centimeters in diameter. These buds would separate from the mother's body and fly away in the form of a fine spray carrying the nuclei of different heavy elements. Meyer and Teller suggest that, whereas the heavy elements must have originated by the budding process, lighter elements must have resulted from some such frozen equilibrium as was previously described. It is difficult, however, to see how the two theories can be amalgamated.

Hypothesis of the "Ylem"

A third possible way in which atomic species might have been formed was proposed a few years ago by the author and developed in some detail in collaboration with his colleagues, Ralph Alpher, R. C. Herman, J. S. Smart, Enrico Fermi, and Anthony Turkevich. This theory occupies an intermediate position between the hypotheses of frozen equilibrium and of the spontaneous break-up of primordial nuclear fluid. The original state of matter is assumed to be a *hot nuclear gas* (not a fluid). It is also assumed that physical conditions at that epoch were changing so rapidly that no real equilibrium was ever established, and the situation must instead be treated in terms of a fast dynamic process. Both assumptions are far from being arbitrary; they follow, in fact, from the general theory of expansion described in Chapter II,[5] which shows that during the early stages of the expanding universe the temperature must have been extremely high and changes exceedingly rapid.

We can illustrate the difference between "frozen equilibrium" and "fast dynamics" by the analogy of a school subjected to certain changes in its educational program. If the program were kept unchanged for a number of years (longer than the time spent in that school by any individual student) and then changed suddenly, the alumni who graduated prior to the change will be in equilibrium with the first program. For example, if the change consisted in the abolition of previously taught classical languages, the earlier alumni will know Greek and Latin. If, however, program changes were being made all the time, the knowledge of graduating students would be different for each graduating class, and in no case would it correspond to any of the different programs which were in force at various times. Such a situation, no doubt very harmful in an educational pro-

[5] No previous author based his assumption on the results of relativistic cosmology.

gram, is quite helpful for the understanding of the origin of atomic species.

Let us now consider the state of matter during the first minutes of the expansion process, when the temperature of the universe was many billions of degrees high.[6] At these temperatures the kinetic energy of thermal motion was measured in millions of electron volts, and the particles were rushing around with velocities comparable to those obtained in modern "atom-smashing" machines. No composite nuclei could have existed under these conditions, and the state of matter must be visualized as a hot gas formed entirely by nuclear particles; that is, protons, neutrons, and electrons. It is known, of course, that free neutrons are intrinsically unstable and break up spontaneously into protons and electrons within 13 minutes after being kicked out of the nucleus. At very high temperatures and pressures, however, free neutrons can co-exist with protons and electrons in quite considerable numbers. In fact, under such extreme conditions a kind of dynamic balance will be established. The progressive decay of neutrons ($n \rightarrow p + \bar{e}$) will be compensated for by the building of new neutrons by the reverse process, collisions of protons and electrons ($p + \bar{e} \rightarrow n$). We will call this primordial mixture of nuclear particles "Ylem," [7] reviving an obsolete noun which, according to Webster's Dictionary, means "the first substance from which the elements were supposed to be formed."

Next we can ask what happened to the Ylem when its density and temperature began to drop as the result of the rapid expansion taking place in the young universe. As soon as the Ylem began to cool, the neutron-producing reaction ($p + \bar{e} \rightarrow n$), which had been supplying fresh neutrons, must have first slowed down and then completely stopped because of the lack of fast thermal electrons. The spontaneous decay of neutrons must then have proceeded without any compensation, so that not

[6] See formula on page 42.
[7] Pronounced: *ī'lĕm.*

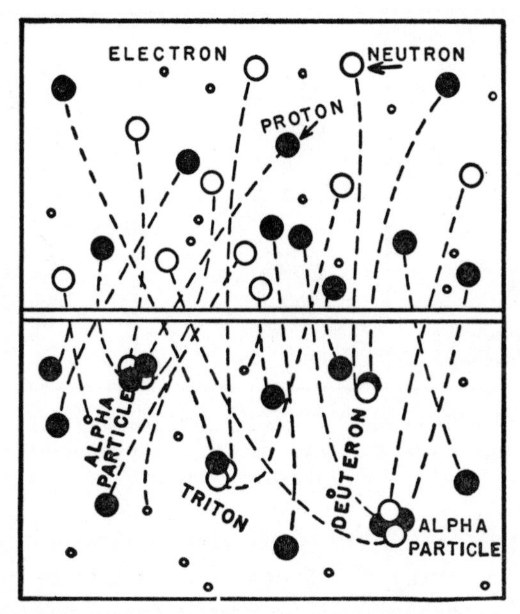

Fig. 10. A schematic picture of the process which took place in the cooling Ylem. Out of ten neutrons originally present (top), three have decayed, four have entered the structure of the two alpha particles, two have been used to build a triton, and one has been used to build a deuteron

very many free neutrons could have been left in the mixture at the end of the first hour after the expansion started. On the other hand, the lowering of the temperature was favorable for the "aggregation process" in which the neutrons still remaining were attaching themselves to protons, thus forming aggregates of particles with different degrees of complexity. These aggregates were the prototypes of the atomic nuclei existing today. The competition between neutron decay and aggregation process is illustrated in Fig. 10.

Nuclear aggregates which originated that way contained many more neutrons than protons originally, because the aggregation of protons was strongly hindered by the repulsion between their like electric charges. Since the stability of atomic nuclei requires about equal numbers of both kinds of particles

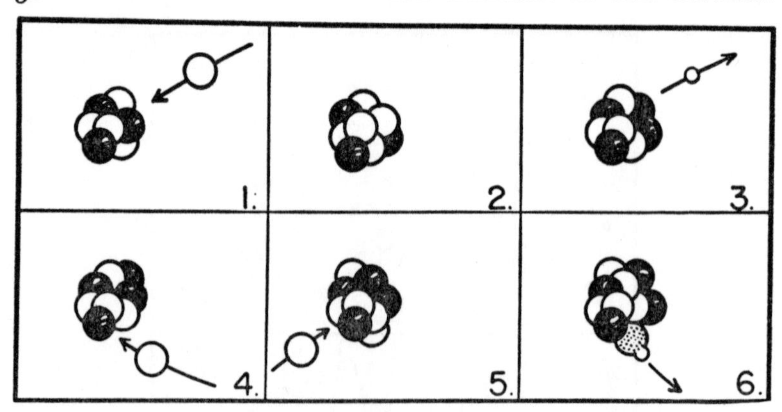

Fig. 11. A schematic picture of intermittent processes of neutron-capture and β-decay. (Black balls indicate protons; white balls, neutrons; smaller balls, negative electrons)

(the stable nucleus of the oxygen atom, for example, consists of eight protons and eight neutrons), the growth of particle-aggregates must have been interrupted from time to time by the processes of β-transformation, in which the excess neutrons were turned into protons by emitting a negative electron. The growth of a complex nucleus by intermittent neutron-captures and β-transformations is shown schematically in Fig. 11.

Since the total time permitted for aggregation was limited to about an hour (after that time no neutrons were left and the temperature of the Ylem dropped below the limit necessary for proton reactions), the results of the process must have depended on the rate at which the aggregation was taking place. This rate was determined by the density of the Ylem, since the higher the density the more collisions between particles must have taken place per unit volume per unit time. If, at the moment when the temperature of the Ylem dropped to the point—below 10^9 degrees centigrade—at which aggregation is possible, the Ylem's density was already very low, most of the neutrons originally present would have broken up into protons and electrons before having a chance to encounter a growing nucleus and attach themselves to it. On the other hand, if at that moment the density of the Ylem was very high, most of the neutrons originally present would have been used up for building heavy nuclei, and only a very few would have decayed into protons and electrons. In the first case, the process would result in lots of hydrogen atoms and practically no heavy elements; in the second, in very little hydrogen and lots of heavy nuclei. We see that, in order to cook the atomic species properly, and in percentages corresponding to those actually found in nature, one must make a rather delicate adjustment of Ylem density (or pressure) during the cooking period. Too much pressure used in this "primodial pressure cooker" would overcook hydrogen and undercook uranium, whereas too little pressure would produce the opposite result. This kind of adjustment must yield the value of the constant in the density expression discussed in Chapter II, which is the only missing link in the chain of conditions determining the initial physical properties of the expanding universe.

The first result of the Ylem theory, namely the conclusion that the entire process of atom-building took less than an hour, might well be greeted with surprise and disbelief. Isn't it silly indeed to talk about something which took place billions of years ago but lasted only for about an hour! However, it must be remembered that in nuclear phenomena the relative time scale is rather elastic. For example, the nuclear chain reaction which takes place in an exploding atomic bomb is over in just a few microseconds, but some radioactive fission products of the exploded bomb can still be detected at the explosion sites several years later. The ratio of several years to several microseconds is

$$\frac{\sim 3 \cdot 10^{7}}{10^{-6}} = 3 \cdot 10^{13};$$

the ratio of 3 billion years to 1 hour is

$$\frac{10^{17}}{3600} = 3 \cdot 10^{13}$$

If we are not surprised that such a short process as an A-bomb explosion can produce radioactive material lasting for several years, why should we be surprised that primordial nuclear processes produced the atoms of uranium, thorium, etc., which are still in existence nowadays, several billion years later.

One may question our right to apply the empirical data of today's nuclear physics to processes which took place under such unusual circumstances as existed during the highly compressed initial stages of the expanding universe with the material heated to many billion degrees. But here again the situation is not without parallels. In fact, as has already been mentioned, at these temperatures the thermal energies of nuclear particles were in the neighborhood of 1 million electron volts, which are the energies used every day in nuclear laboratories for the study of nuclear reactions. It is immaterial whether the colliding nuclear particles have these high energies as the result of thermal motion or because they were accelerated in special high-tension machines. We will also learn that the density of matter during the period of atom-cooking was comparable to that of atmospheric air, so that there is no reason to doubt the applicability of physical laws pertaining to gases or of the laws of gravity which have governed the process of expansion. One finds that, in fact, there is very little choice in assumptions on which to base our calculations. All we have to do is to accept relativistic formulae for the universal expansion, and empirical data concerning various nuclear reactions, and see if the calculations lead to a result which resembles the observed abundances of known atomic species.

Theoretical abundance curve

The mathematical equations that describe the building process just discussed are very simple indeed. They state that the rate of change of the number of nuclei with a given atomic weight is equal to the difference between the rate of production by neutron capture from the next lower atomic weight group and the rate of disappearance through neutron capture by the next higher weight group.[8]

The problem we are facing here, although much more difficult, is very similar to that of heat conduction along a thermally insulated bar heated at one end. In this problem too, the increase of temperature in any section of the bar is the difference between the incoming heat from the left and the outgoing heat to the right. And, by an amusing coincidence, the coefficient of heat conduction in the bar problem is usually denoted by the same Greek letter σ which designates the capture cross section of neutrons in our nuclear problem. This analogy is represented in Fig. 12. If the bar consists of homogeneous material (say, iron), so that the heat conduction coefficient σ is the same all along its length, temperature distribution along the bar at various times after the heating started will be represented by exponential curves as shown. If the probabilities of neutron cap-

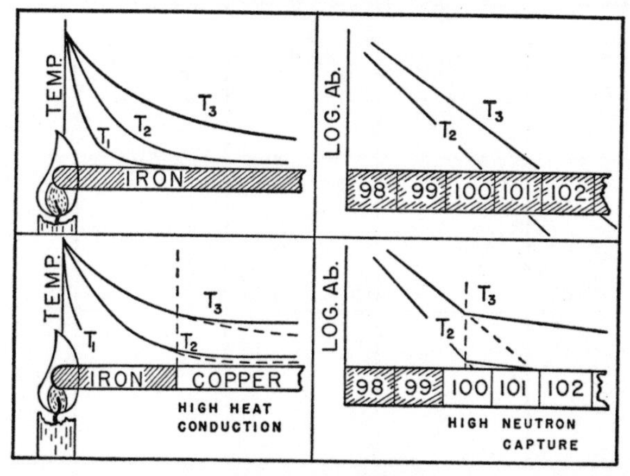

Fig. 12. Analogy between heat conduction and the nuclear building-up process

ture (capture cross sections σ) were the same for the nuclei of all atomic weights, the building process would also cause the

[8] See Appendix, page 143.

abundances to decrease exponentially with atomic weight. This would explain the descending part of the empirical abundance curve (Fig. 6), but not the almost horizontal part in the region of the heavier elements. It is easy to see, however, that the flattening of the curve can be obtained if one assumes that σ increases from left to right. Indeed, if we consider a metal bar made half of iron and half of copper (which is a better heat conductor than iron), with the iron end in the flame, we will expect to have a steep temperature gradient in the iron part, and much more even temperature distribution in the copper part where the heat flows much more easily. Similarly we could expect to obtain the flat run of the abundance curve in the region of the

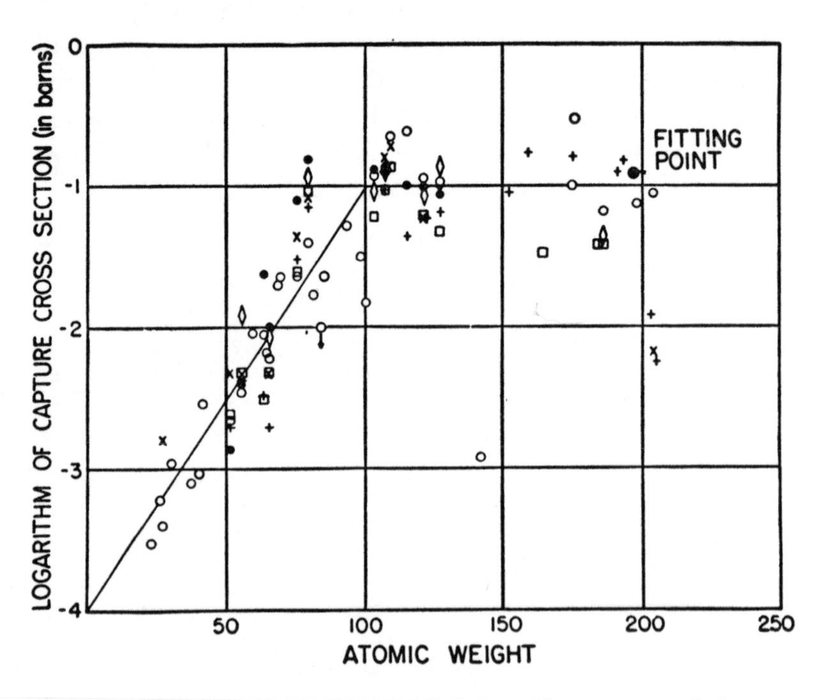

Fig. 13. Capture cross sections, or probability of being captured, for neutrons of about 1-million-electron-volt energy, plotted against the atomic weights of the elements doing the capturing. The continuous line is an average accepted as a basis for further calculations

heavy elements, if the probability of neutron capture σ were greater for heavier nuclei than for light ones. And, lo and behold, this is actually the case, as is shown in Fig. 13, which represents actually measured neutron capture cross sections in various elements for neutron energies in the neighborhood of 1 million electron volts (corresponding to 0.1 billion degrees). The theory of atom-building based on the process of successive neutron capture thereby promises to give a correct explanation of the empirical abundance curve.

Before considering the results of integration of the equations for such atom-building, we want to indicate a purely empirical relationship which exists between the neutron capture cross sections shown in Fig. 13, and the abundances of atomic species shown in Fig. 6. If, combining these two curves, we plot the logarithms of the directly observed abundances against the measured cross sections, we get the result indicated by the small

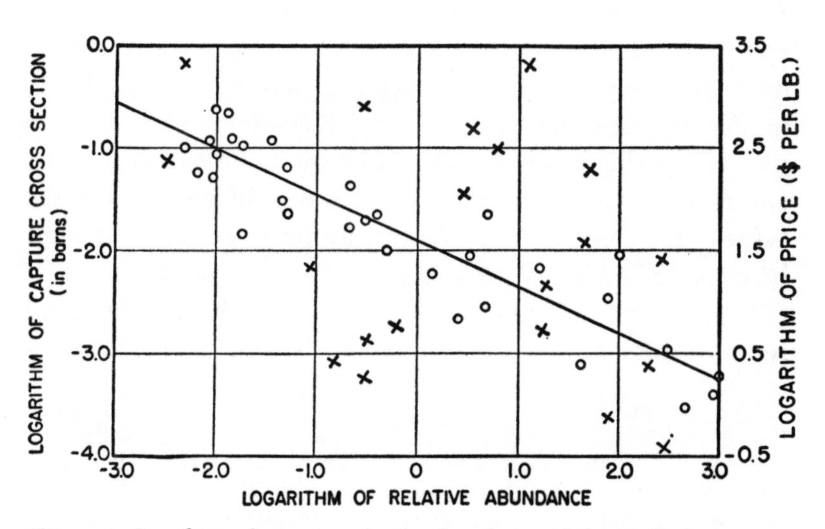

Fig. 14. Correlation between relative abundance of chemical elements on one side, and capture cross sections (circles) and commercial price per pound (crosses) on the other. There is a very good correlation between abundance and capture cross sections; a very poor one between abundance and price

circles in Fig. 14. We can see that the circles fall very nicely on a straight line, indicating a simple relationship between the two quantities. The crosses in Fig. 14 indicate a comparison of the known relative abundances of various chemical elements with their market prices as given in the catalogue of a large chemical firm. In this case the points are scattered far more irregularly through the entire field, indicating that the market prices of pure elements are determined not so much by their abundance or variety in nature as by their industrial usefulness and the existing facilities for mining and purifying them.

Actual integration of the equations for atom-building was first performed by Ralph Alpher.[9]

The close fit of the calculated curve and the observed abundances is shown in Fig. 15, which represents the results of later calculations carried out on the electronic computer of the National Bureau of Standards by Ralph Alpher and R. C. Herman (who stubbornly refuses to change his name to Delter). It is apparent that the two upper curves, marked $10\rho_{ST}$ and $7\rho_{ST}$, overcook heavy elements, whereas the curve marked $1\rho_{ST}$ badly undercooks them. A reasonably good fit is achieved by the curve $5\rho_{ST}$ and probably still better results would be obtained if the curve $4\rho_{ST}$ were calculated. To get the satisfactory curve the value of the constant in the density formula of Chapter II must be taken to be $1.2 \cdot 10^{-3}$ grams per cubic centimeter.

We may note here that, in addition to providing a rather good representation of the general trend of the empirical abundance

[9] The results of these calculations were first announced in a letter to *The Physical Review*, April 1, 1948. This was signed Alpher, Bethe, and Gamow, and is often referred to as the "alphabetical article." It seemed unfair to the Greek alphabet to have the article signed by Alpher and Gamow only, and so the name of Dr. Hans A. Bethe (*in absentia*) was inserted in preparing the manuscript for print. Dr. Bethe, who received a copy of the manuscript, did not object, and, as a matter of fact, was quite helpful in subsequent discussions. There was, however, a rumor that later, when the α, β, γ theory went temporarily on the rocks, Dr. Bethe seriously considered changing his name to Zacharias.

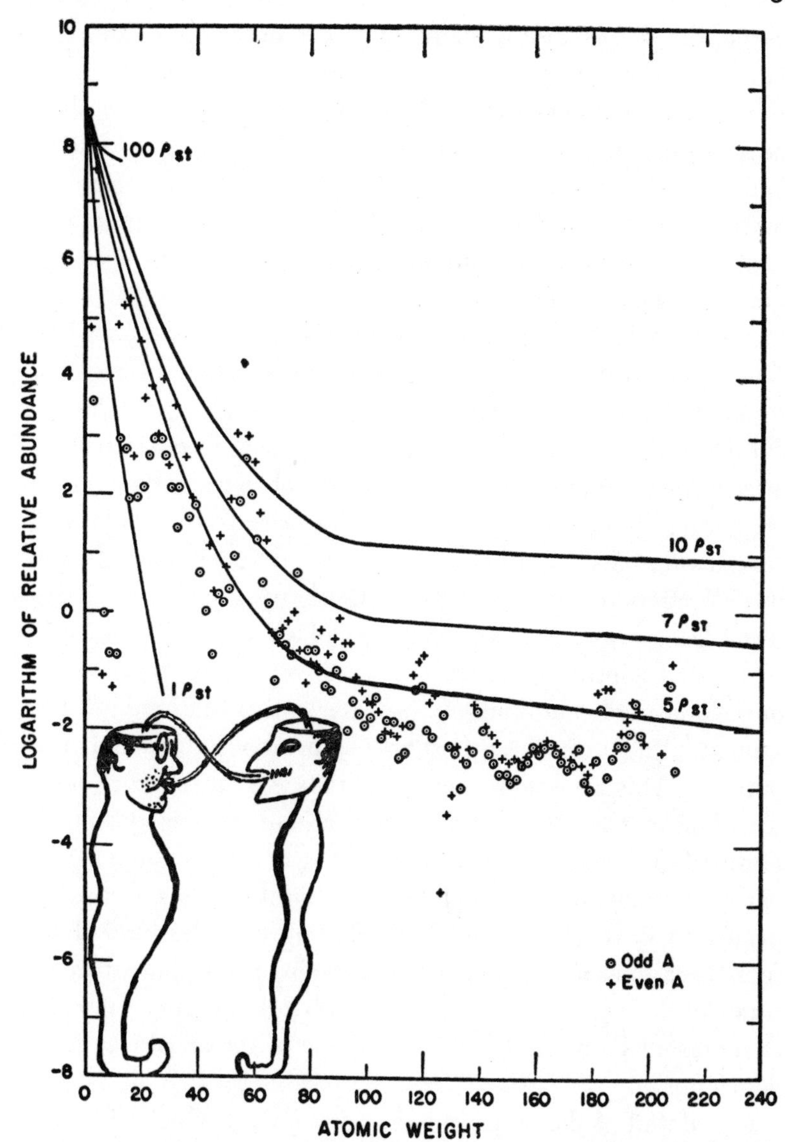

Fig. 15. Calculated abundance curves, after Ralph Alpher (left) and R. C. Herman (right). Empirical points are the same as those in Fig. 6

curve, the theory also explains some characteristic details pertaining to the abundance of certain elements. A few groups of elements have abnormally high abundances, thus causing sharp peaks to rise above the generally smooth curve in Fig. 6. If we look into the situation more carefully we find that these abnormally high abundances (with the single exception of a sharp peak near iron) correspond to nuclear species with a "magic number" of either neutrons or protons. In nuclear physics "magic numbers," which are 2, 8, 20, 50, 82, 126, play the same role as do the numbers 2, 8, 18, etc., in ordinary chemistry. They represent the number of particles in the nucleus which forms a complete shell in its internal structure. Just as in chemistry the atoms with complete electronic shells (helium, argon, neon, etc.) are self-consistent and chemically inert, so the nuclei with "magic numbers" of either protons or neutrons are considerably less effective in capturing new particles; we say that their capture cross sections are abnormally small. Obviously, nuclei with such small capture cross sections must represent "bottlenecks" in the continuous building-up process, so that material must be accumulating near these bottlenecks in abnormally high proportions. This argument seems to provide a satisfactory interpretation of the observed abnormal abundances in the neighborhood of "magic-number" nuclei, but it is only fair to point out that nuclei with completed structural shells are also expected to have abnormally high binding energies, and therefore the abnormal abundances could also be explained in the frame of the equilibrium theory, provided that theory were able to represent correctly the general shape of the empirical abundance curve.

One detail of the empirical abundance data which cannot as yet be satisfactorily explained by the building-up process is the existence of the so-called "shielded isotopes." These are stable nuclear species which could not have been formed by several successive β-transformations from the originally radioactive nuclei, because some of these transformations would require larger amounts of energy than are available in the nucleus. Since, according to standard Ylem theory, all nuclei are first built with an excess of neutrons and brought into a normal state only later by a series of successive β-transformations, such shielded isotopes should not be produced at all. However, shielded isotopes do exist in nature. There seems to be a possibility that they could be produced by the so-called "n, 2n process," in which a fast incoming neutron produced in some previous reaction knocks two neutrons from the nucleus with which it collides. However, this problem needs further study.

The case of light elements

The calculations just described give us an over-all picture of atom-building by the process of successive neutron capture. In order to formulate the problem in such a way as to bring it within the reach of mathematical analysis, the empirical curve of capture cross sections was smoothed out, and no processes except that of direct neutron capture were taken into consideration. Another method of attack would be to consider the lightest nuclei only but to perform the calculations with every possible detail. This was first attempted by the author, and later with considerably more detail by Enrico Fermi and Anthony Turkevich. These writers give a "blow-by-blow" description of what must have happened to the six simplest nuclei (neutrons, protons, deuterons, tritons, tralpha-particles, and alpha-particles) during the first 35 minutes of the expansion of the Ylem. Due to the low atomic numbers of these nuclei, possible reactions would also involve those between charged particles, so that altogether twenty-eight different types of nuclear processes have to be considered. The results of the calculations, made for an initial density constant of 10^{-3} grams per cubic centimeter and assuming a 100-per-cent neutron content at the beginning, are shown in Fig. 16. We see that during the first 5 minutes

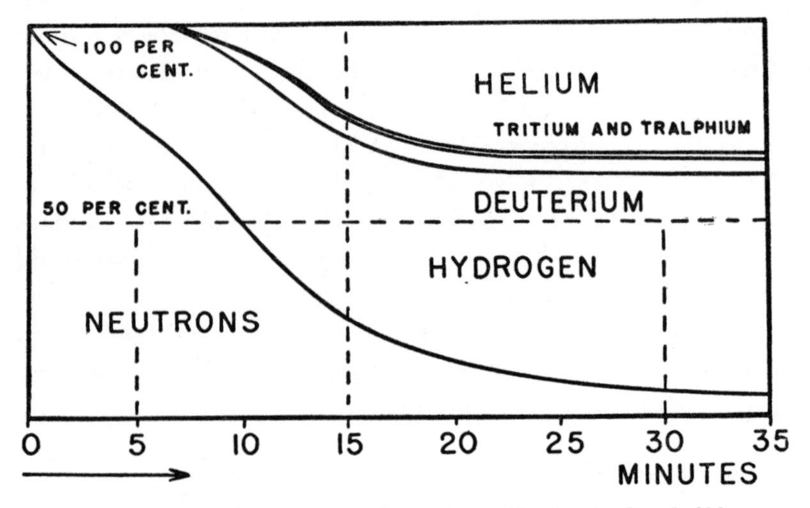

Fig. 16. Chemical changes in the universe during its first half-hour (after Enrico Fermi and Anthony Turkevich)

the temperature of the universe was still too high to permit the formation of complex nuclei, so the only nuclear process going on during that period was the spontaneous decay of neutrons into protons and electrons. With further lowering of the temperature, neutrons and protons began to stick together, forming deuterium nuclei, which, however, never accumulated in large quantities because they were rapidly transformed into ordinary helium. Tritium (H^3) and tralphium (He^3), serving as transition stages from mass 2 to mass 4, were always present in extremely small quantities, as is shown in the diagram. We see that toward the end of the first 30 minutes slightly more than half of all the original Ylem became hydrogen and slightly less than half of it became helium. This is exactly the relationship one finds in the universe today. We also see that by that time the original neutrons had almost completely decayed, and deuterium constituted about 1 per cent of the entire mass. The latter conclusion does not, of course, correspond to physical reality (deuterium is actually exceedingly rare in nature), and resulted from the fact that these calculations did not follow the building-up process beyond helium. In reality, most of the deuterium formed in this way must have been consumed in the process of building heavier elements.

In spite of their initial success, Fermi and Turkevich ran into serious difficulty when they attempted to carry their calculations of the building-up process beyond helium. The trouble lies in the fact that the nucleus of mass 5, which would be the next stepping stone, is not available. Due to some peculiar interplay of nuclear forces, neither a single proton nor a single neutron can be rigidly attached to the helium nucleus, so that the next stable nucleus is that of mass 6 (the lighter isotope of lithium), which contains two extra particles. On the other hand, under the assumed physical conditions, the probability that two particles will be captured simultaneously by a helium nucleus is negligibly small, and the building-up process seems to be stopped short at that point. Several ideas were suggested as to how that crevasse could be jumped in order to continue our climb up the periodic system of the elements. In their original studies Fermi and Turkevich considered the reaction

$$He^4 + T^3 \rightarrow Li^7 + \gamma \text{ quantum}$$

which does not involve an intermediate nucleus of mass 5. They found, however, that, under the existing conditions of density and temperature, this reaction seems to be too slow to provide a sufficient quantity of heavier nuclei. It could be considerably accelerated by the presence of strong resonance (in the neighborhood of 400 kilo electron volts), but no indications of such resonance have so far been found in experimental studies of the Li^7 nucleus.

Another ingenious method of crossing the mass 5 crevasse was proposed by E. Wigner. It is known as the method of the "nuclear chain bridge." Wigner's plan is illustrated in Fig. 17, which shows that all that is required for building a chain bridge is an assumption that there was originally one single nucleus

Fig. 17. Wigner's proposal as to how to jump across the mass 5 crevasse

on the right-hand side of the crevasse. Such an assumption can easily be granted, since some building up is still maintained across the crevasse by the reaction

$$He^4 + T^3 \rightarrow Li^7 + \gamma \text{ quantum}$$

in spite of the low probability of its occurrence. This single nucleus on the right-hand side of the crevasse (C^{11} in the picture) may unite with another nucleus on the left-hand side (T^3 in the picture), giving rise to two lighter nuclei (Li^6 and Be^7), which are both on the right-hand side. By the regular process of neutron capture these two nuclei may be built up into a pair of nuclei identical with the originally assumed nucleus (C^{11}), and the process begins again with double volume. After a number of such branchings, a large number of nuclei will be built up to participate in the process and the mass transfer across the crevasse will be running at normal high speed. Unfortunately, the particular reaction used in Fig. 17 to illustrate Wigner's idea is not quite suitable for the task, so that some other similar reaction should be found to take its place. So far no such reaction has been found, which may be due simply to a deficiency in our information about the various isotopes which may be involved.

It is also quite possible that the difficulty of the mass 5 crevasse can be removed without any special devices, by simply considering the thermonuclear reactions in somewhat more detail than has been done so far. In all previous calculations it has been tacitly assumed that the temperature of nuclear gas always remains equal to the temperature of radiation as given in the formula at the end of Chapter II. However, all the nuclear reactions involved in the building-up process are connected with the liberation of considerable amounts of nuclear energy. This opens up the possibility that *the reacting nuclear gas may become hotter than the radiation in which it is immersed.* In fact, while the temperature of radiation is dropping, the temperature of nuclear gas may be rising, reaching rather high values before it finally begins to go down too. This temporary rise in the temperature of the gas will not greatly affect the results of Alpher's and Herman's calculations shown in Fig. 15, since the probability of neutron-capture by heavier nuclei is not very sensitive to temperature. However, the rise in gas temperature will accelerate quite considerably all thermonuclear reactions between light elements and may change quite radically the results of the calculations made by Fermi and Turkevich. In particular, this higher gas temperature will strongly favor the reaction

$$He^4 + T^3 \rightarrow Li^7 + \gamma$$

and it may result in the production of adequate amounts of lithium and other heavier elements.

Unfortunately, the calculations involving the self-heating of the reacting nuclear gas are tremendously complicated and can be carried out only by means of modern electronic computing machines.

To sum up these somewhat lengthy discussions, we may say that even though the Ylem theory of the origin of atomic species is by no means complete, it still provides a reasonably satisfactory picture of what may have happened during the early stages in the history of our universe.

Could heavy elements be formed in stars?

Another possibility considered by Hoyle and others is that heavy elements were produced in the hot interior of stars. As we see in Chapter V, energy production in normal stars, such as our sun or Sirius, is due to a slow thermonuclear process in which hydrogen gradually transforms into helium. A temperature of about 20 million degrees, which causes this nuclear reaction, is, however, not high enough to induce the reactions between heavier nuclei and to produce heavy elements in any appreciable amounts. But when a star exhausts its hydrogen, the main source of stellar energy, it goes through a series of states in which its central temperature and density rise to considerably higher values. E. Salpeter has shown that under such circumstances more complicated reactions are taking place, such as:

$$_2He^4 + {_2He^4} + {_2He^4} \rightarrow {_6C^{12}}$$
$$_6C^{12} + {_2He^4} \rightarrow {_8O^{16}}$$
$$_8O^{16} + {_2He^4} \rightarrow {_{10}Ne^{20}}$$

This gradual building up of heavier and heavier elements proceeds all the way to iron, the nuclei of which possess the greatest stability. The gap between iron and uranium is more difficult to cover and requires the presence of neutrons which must be produced in the reactions between the lighter elements. At the end of their evolution stars become internally unstable and explode violently (supernovae explosions), scattering their material throughout interstellar space. This material can later condense again, forming "second-hand" stars with large amounts of heavy elements. According to this point of view, our sun is such a second-hand star, made up from the material scattered by one of several supernovae which exploded billions of years ago.

It is possible that both the Ylem theory, described earlier in this chapter, and the theory according to which heavy elements are produced in aging and exploding stars play equally important roles in the life of the universe. As mentioned in Chapter II and again in Chapter IV, W. Baade found two classes of stellar population: Type I, found in spiral arms and associated with a large amount of interstellar gas and dust, and Type II, found in the central bodies of the galaxies, which seems to be formed exclusively of stars with no interstellar material whatsoever.

Spectroscopic studies carried out by M. Schwartzschild and others indicate that the stars of Type I (to which our sun belongs) contain heavy elements in the same proportions as these are found within the solar system. On the other hand, the stars of Type II are almost entirely deprived of heavy elements, having an amount at least 100 times smaller than that in stars of Type I. Thus one might think of the stars of Type II as formed from the original material produced during the era of the Big Squeeze, while stars of Type I have been enriched by heavy elements in the processes of supernovae explosions. Only time will tell whether this corresponds to the truth.

CHAPTER IV

The Hierarchy of Condensations

First clouds

After the full complement of the atomic species had been formed during the first hour of expansion nothing of particular interest happened for the next 30 million years. The hot gas consisting of the newly formed atoms continued to expand, and gradually its temperature became lower and lower. When the temperature had fallen from the original billions of degrees to only a few thousand degrees, that portion of the gas formed by vapors of various elements with high melting points condensed into fine dust, which continued to float in the prevailing mixture of hydrogen and helium. This dusty gas mixture (1 milligram of gas and several micrograms of dust per million cubic kilometers of space) still exists in interstellar space, causing the so-called interstellar absorption lines and the reddening of the faraway stars. Sometimes this interstellar material accumulates into giant clouds of irregular shapes. Such clouds are known as luminous nebulae or dark nebulae depending on whether or not they are illuminated by the nearby stars (Plates IV and V). If this state of affairs had continued indefinitely, the universe of today would contain nothing but this highly diluted gas-dust mixture, with its temperature close to absolute zero. We know, however, that at present the matter of the universe is highly differentiated, forming galaxies, stars, and planets. When and why did this differentiation take place? The answer to that problem lies in the relation between the radiation densities and gas densities discussed in the preceding two chapters.

We have seen that during the early stages of expansion the mass-density of radiation contained in any volume of space exceeded the density of ordinary matter by a very large factor. In that period radiant energy was the ruling agent of the evolution of the universe; the atoms were easily kicked about by their collisions with the powerful light quanta. Since, by its nature, the radiation must have filled space uniformly, the distribution of ordinary matter also must have been quite uniform. But, as we have also seen, the expansion of the universe was gradually stacking the cards in favor of matter, and a time must have come when the mass-density of radiant energy fell below the density of ordinary matter. From then on, matter took over the leading role in the evolutionary process and it is logical to assume that the present highly differentiated state of the material universe is the result of this change of leadership. There is hardly any doubt that the primary role in the events that occurred when matter took over was played by the force of Newtonian gravity, acting between the material particles scattered nearly uniformly through space. As was shown by the famous British astronomer Sir James Jeans almost half a century ago, a gas subjected to gravitational forces and filling an unlimited space is intrinsically unstable and is bound to break up into separate giant gas clouds (Fig. 18). This instability results from the fact that embryonic local condensations which occurred in that gas from purely accidental causes would be kept from dissolving again by the action of gravitational forces. Such occasional condensations occur on a minor scale in ordinary atmospheric air, but the forces of gravity are too weak to hold them together. For larger gas masses, however, the forces of gravity become increasingly important, and the large-scale condensations which may form in unlimited space will not be able to resolve themselves again. As a result, the gas breaks up into large individual clouds with almost complete vacuums between them.

The size of these condensations is determined by the condi-

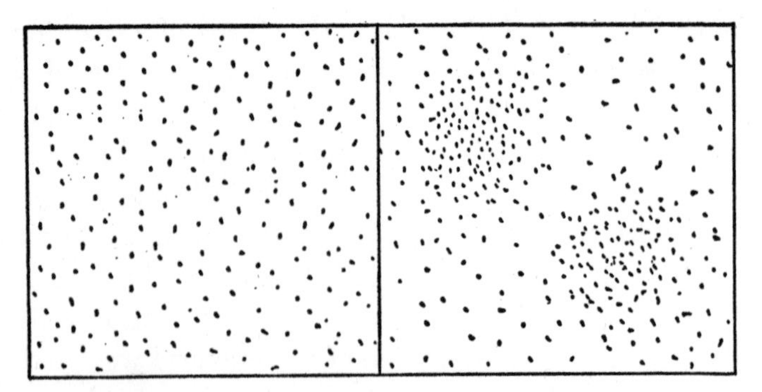

Fig. 18. Condensation of points in an originally uniform distribution

tion that the gravitational potential on their surface must be greater than the energy of thermal motion of the gas particles, so that, once the condensation is formed, the particles which are part of it cannot escape its field of gravity. This is the same important concept of "escape velocity" that applied earlier to the problem of unlimited expansion of the system of galaxies. It is easy to calculate [1] the radii and masses of such condensations in an originally uniform gas of the density and temperature given by the expressions:

$$[\text{radius}] \geqq 4.4 \cdot 10^{7} \sqrt{\frac{[\text{temperature}]}{[\text{density}]}} \text{ (centimeters)}$$

and

$$[\text{mass}] \geqq 3.9 \cdot 10^{22} \sqrt{\frac{[\text{temperature}]^{3}}{[\text{density}]}} \text{ (grams)}$$

If, as an example, we apply these formulae to atmospheric air by using a temperature of 300 degrees absolute and a density of 10^{-3} grams per cubic centimeter, we will obtain for the radius of the condensations the value

$$2 \cdot 10^{10} \text{ cm} = 2 \cdot 10^{5} \text{ km}$$

which is larger than the diameter of the earth! Therefore our

[1] See Appendix, pages 142–43.

atmospheric air does not break up into separate air balls simply because the layer of atmosphere is too thin.

There were no such limitations on the primordial gas filling the unlimited space of the expanding universe, and we can calculate the sizes and masses of condensations which could form during various stages of its evolution. Using the formulae for time changes of temperature and of density in the expanding universe,[2] and substituting them in the expression for mass, we find that the time factor is canceled out. Thus the mass of primordial gas clouds formed in a gravitational break-up process of originally uniform material would always be the same no matter at what stage of the expansion the condensation process took place. Substituting numerical values, we find that *the smallest possible mass of these condensations is* 10^{40} *grams*, which exceeds the mass of our sun several million times. Although the estimated minimum masses of primordial gas clouds fall short of the values usually accepted for the masses of individual galaxies, the result is still very gratifying. Indeed, since the values for density and temperature used in our calculations were derived on the basis of purely nuclear data, we may have here a bridge (or rather a viaduct) uniting the microcosmos of nuclear particles with the macrocosmos of stellar systems!

Several reasons can be cited as to why the calculated value for the mass of condensations falls short of the observed galactic masses. First of all, our formula gives us only the minimum mass, and actual condensations may easily have been much larger. A second and more important reason lies in the fact that Jeans' original formula, used in our estimate, is strictly applicable only in the case of a non-expanding gas. If, as was really the case, expansion is present, allowance must be made for the kinetic energy of expanding gas masses, and the minimum mass which could keep itself together by gravity must obviously be considerably greater. These problems, and others, are

[2] See Appendix, formulae (5) and (6), page 143.

awaiting further and more detailed studies for their solution.

While the masses of primordial gas clouds resulting from the break-up of the originally uniform distribution of matter turn out to be independent of the epoch of their formation, their geometrical dimensions certainly depend upon it. In fact, clouds formed during the early stages of expansion would have to be rather small in size and quite dense, while clouds formed later would be much larger and rather dilute. The observed fact that different galaxies, while not identical, do not show appreciable differences in their diameters indicates that they all must have been formed at about the same time. And it is only logical to assume that the time of their formation coincided with the time when matter succeeded radiant energy as the decisive factor, and the force of Newtonian gravity became of primary importance. We can obtain the date of that event by assuming that, once the clouds were formed, their density remained unchanged and only their distances from one another continued to increase with time. Since we know from observational evidence that the average distances between neighboring galaxies are about a hundred times greater than their mean diameters, we must place the separation era at about one-hundredth of the present age of the universe, at about the date when the universe was some 30 million years old. At that time the mean density of matter in the universe must have been equal to the present mean density within individual galaxies, which is of the order of 10^{-24} grams per cubic centimeter. On the other hand, using the expression for temperature variation given in Chapter III, we find that at the age of 30 million years the universe had a temperature of about 300 degrees absolute, so that the mass density of radiation was also 10^{-24} grams per cubic centimeter.[3] This result confirms our guess that the formation of the first clouds

[3] In fact:

$$\frac{aT^4}{c^2} = \frac{10^{-14} \cdot (300)^4}{10^{21}} \cong 10^{-24}\ \text{g/cm}^3$$

took place at the time when the mass density of radiant energy was falling below the density of ordinary matter.

If we could get H. G. Wells' "time machine" and go back to the year 30,000,000 A. C. (After Creation), we would find ourselves floating in an almost complete vacuum, comparable to that which exists today in the space between the stars inside our galaxy. It would be pitch dark around us, since the brilliance of the first days of creation (comparable to that of the center of an exploding atomic bomb) had by that time been completely dimmed by the expansion process, and the stars which illuminate the universe today had not yet been formed. We would, however, be comfortably warm, since the prevailing mean temperature of about 300 degrees absolute is close to what we call room temperature!

Before leaving the subject of the formation of gaseous protogalaxies, we must mention another factor which undoubtedly played an important role in their formation. As we have seen (Chapter II), the potential energy of Newtonian gravity between the galaxies today amounts to only about 1 per cent of their kinetic energy of motion. In other words, the galaxies are today completely disengaged from their mutual gravitational attraction. However, looking back in time, we find that at the time of galactic separation their mutual distances were only 1 per cent of what they are now and consequently their mutual gravitational energy must have been a hundred times as great. At that time the mutual recession of galaxies was still strongly hampered by their mutual gravitational attraction. This situation is similar to that of a rocket which has more than escape velocity but at the instant under consideration is still "climbing up" through the gravitational field of the earth and is losing velocity in the process.

Thus *the original break-up of the uniformly expanding material of the universe took place at the time when this material ceased to be "gravitationally coherent."* Whatever the exact theory of galactic separation may turn out to be, there is hardly any doubt that this phenomenon must have been closely connected with the disappearance of gravitational coherence between the expanding masses, as well as with radiant energy's loss of the principal role.

Galactic rotation and turbulence

When a bulk of continuous material breaks up violently into many fragments, the pieces fly apart, spinning rapidly, as do the fragments of an artillery shell exploding in midair. On the basis of general mechanical considerations, one has to expect that the available energy will be more or less evenly distributed between the translatory and rotational motion of the fragments. And, indeed, as we saw in the beginning of Chapter II, rotational and translatory energies of galaxies are of the same order of magnitude.

Depending on the degree of rotation obtained by various proto-galaxies in the separation process, their gaseous bodies

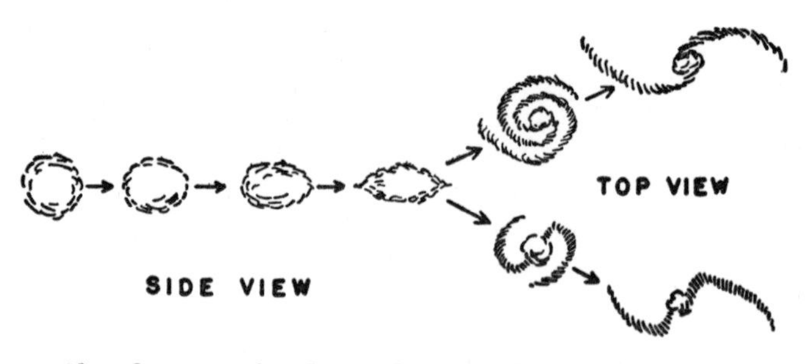

Fig. 19. Classification of galactic forms: spherical, elliptical, ordinary spiral, barred spiral

must have assumed different shapes. The few that by pure chance received only negligible amounts of rotation assumed nearly spherical shapes. Others assumed ellipsoidal shapes with varying degrees of elongation depending on their rotational speeds (Fig. 19). However, most of these originally gaseous fragments had such high rotational speeds that their bodies were flattened out into the shape of a lens and material started streaming out from the sharp edge, to form the familiar picture of spiral arms. In spite of a large amount of work done on this problem our understanding of the various shapes of spiral arms and the details of their origin is still far from complete. Recent studies, however, indicate that spiral arms play a considerably less important role in the general structure of galaxies than it would appear at first sight. The main bulk of the galactic disk seems to be formed by a multitude of stars rotating on regular

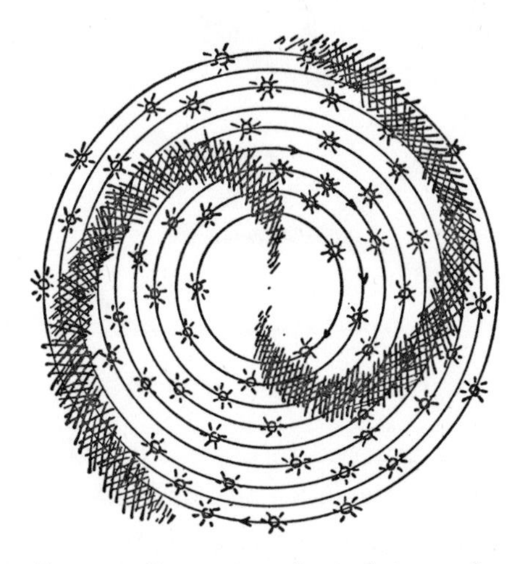

Fig. 20. Formation of spirals in a galaxy. Circles represent the movement of stars; shaded areas gas and dust

circular orbits around the center of the galaxy, whereas the arms are formed by highly diluted dusty gas streamers which are caught by the general rotation of the system and twisted into the shape of spirals (Fig. 20).

The original proto-galaxies were formed entirely of cool gas, with no stars yet present. How did stars come to be? To answer this question we may start with the observation that the rotation of the masses of gas that were the original proto-galaxies

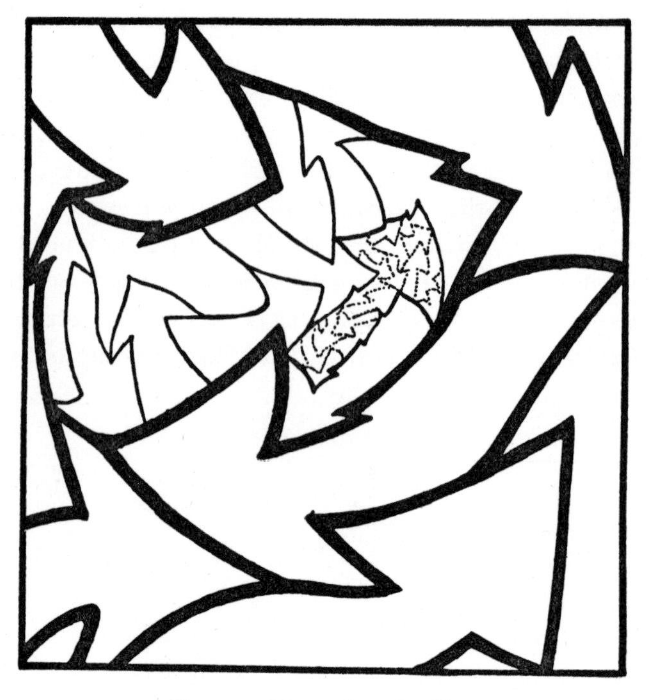

Fig. 21. Hierarchy of eddies

could not possibly proceed in a smooth and uniform manner. The gas masses near the rim must have had a tendency to rotate with lesser angular velocity, a longer rotation period, than the inner ones. In our planetary system the rotation periods increase with the distance from the sun in the center, ranging from about 3 months for Mercury to 165 years for Neptune. But, whereas the planets are separated by a lot of empty space and do not hamper one another's motion, the situation in a rotating gaseous disk will be more like that of a swift river going around a bend. The regular (laminar) flow of water is bound to break up into a multitude of small-scale irregular motions carried along by the main stream. This irregularity of motion, known as turbulence, is a very important factor in all fields of fluid dynamics, from the design of an airplane wing to the explanation of the origin of stars and planets. We can see turbulent motion on the surface of a fast-flowing river when we look down from a bridge; similarly we can feel it in the form of discontinuous gusts of wind blowing in our faces. Turbulence in a fluid medium is a very complicated and completely orderless motion, which is not easy to represent by a simple picture or a diagram. Probably the nearest representation of what turbulence actually is, is given by Fig. 21, in which individual turbulent streamers are represented schematically by arrows of different sizes. If we look at this diagram from a distance we first observe large arrows winding around one another. Closer inspection will show, however, how the large arrows consist of a large number of smaller ones which in turn consist of arrows of still smaller size. Extend this picture in two directions, so that the largest arrows will be almost as large as the entire volume of the fluid, and the smallest ones almost as small as the individual molecules, and you will have a fairly clear picture of what turbulent motion actually is. The verse by L. F. Richardson neatly describes this:

> Big whirls have little whirls,
> That feed on their velocity;
> And little whirls have lesser whirls,
> And so on to viscosity.

In order to draw this at all, the arrows in each category are shown in Fig. 21 as about the same size. In actuality, the hierarchy of turbulent streamers includes all sizes of arrows as well as all directions of motion. Turbulent motion within a fluid includes the rotational as well as the translatory motions of its different elements, and it is because of the rotational type of motion that turbulent streamers are commonly known as "eddies." It would be incorrect to think that turbulent eddies maintain their individuality for a long time, enabling us to represent turbulent motion by giving the position of different eddies for properly chosen intervals of time. Actually, the lives of individual eddies are very short; they usually disappear after traveling a distance comparable to their own diameter, giving rise to new eddies which may move in an entirely different direction. The presence of turbulent motion in a moving fluid leads to an increase in its internal friction known as turbulent viscosity. If, for example, the propeller of a motorboat were to produce nothing but a laminar flow behind its stern, the boat would hardly be pushed at all and would move at a snail's pace. The same is true for the propeller of an airplane; the wings on the other hand should be designed to avoid turbulent motion at their surface, permitting them to slide through the air with the least possible resistance.

At first glance, it would seem impossible to develop a consistent theory for such complex and irregular motion as that presented by a turbulent flow, and until quite recently the study of turbulence was carried out (mostly for engineering purposes) on a purely empirical basis. But during recent years the theory of turbulence has been put on a strict mathematical basis, mostly by the work of G. I. Taylor in England, Theodore von Kármán in the United States, A. N. Kolmogoroff in Russia, and Werner Heisenberg in Germany. One of the main results of that research was the derivation of the so-called "energy spectrum" of turbulent motion. The motion of eddies involves large amounts of kinetic energy, which is being continuously transferred from larger eddies to smaller ones, all the way down from large-scale motion of the fluid to the molecular motion of its constituent particles. We know that in all kinds of friction the kinetic energy of motion changes to heat; in the particular case of turbulent friction this transfer goes down along the hierarchy of eddies of ever-decreasing size. The question we may ask here is: how much energy is stored in eddies of different size (what is their "energy spectrum"?) or, in other words, what are the velocities of the different irregular streamers that constitute turbulent motion? Theoretical considerations, too involved to be even hinted at here, lead to the conclusion that velocity distribution between the eddies of different sizes is governed by the so-called Kolmogoroff law, which can be written as:

$$[\text{velocity}] \sim \sqrt[3]{[\text{size}]}$$

The smaller the size of an eddy, the smaller the corresponding velocity. Imagine, as an example, a fast turbulent stream of water in a channel 10 meters wide. The eddies comparable in size with the width of the channel will have velocities comparable to the total velocity of flow. Eddies 1 centimeter in diameter will have about one-tenth of that velocity ($\sqrt[3]{1000}$), and eddies which are only 10 microns across will have one-hundredth of the velocity of the main flow.

We come now to a very important problem, namely, that of the conditions under which a regular laminar flow of fluid will break up into turbulent eddies. These conditions were established purely empirically by the British physicist Osborne Reynolds, who studied the flow of various liquids at different speeds through tubes of different diameters. He found that the smooth laminar flow of a liquid always breaks up into turbulent eddies when the velocity of flow exceeds a certain limit, a limit which is lower for wider tubes and for liquids of lower viscosity. These findings, which are illustrated in Fig. 22, can be summarized by means of a simple empirical formula:

$$\frac{[\text{density of fluid}] \cdot [\text{velocity of stream}] \cdot [\text{width of the stream}]}{\text{viscosity of fluid}} = R$$

where the dimensionless quantity R is known as the *Reynolds number*. If the density and viscosity of the fluid, and the velocity and width of the flow are such that the number defined by the above equation is less than about 1000, the flow will remain laminar; if the number is greater than 1000, turbulent motion

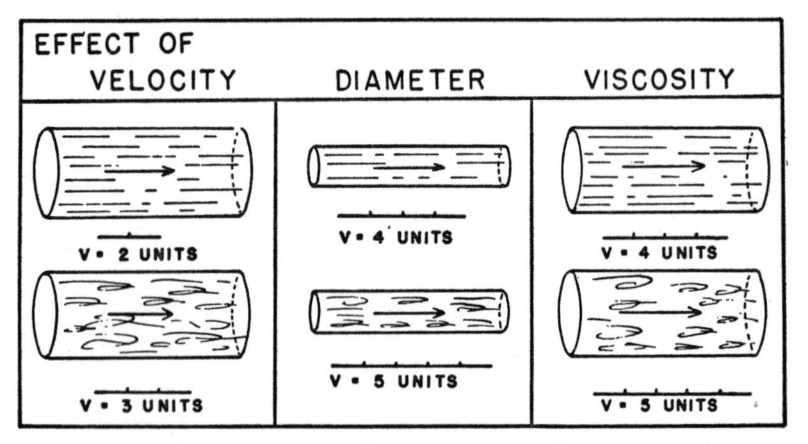

Fig. 22. Break-up of a smooth laminar flow through a tube into turbulent eddies. For a given diameter and viscosity, turbulent motion will appear when the velocity exceeds a certain limit (left). If the diameter is smaller (center), or the viscosity higher (right), the appearance of turbulence will be delayed until higher velocities are reached

will appear. The empirical condition for the appearance of turbulent motion obtained by Reynolds was given a theoretical foundation in the recent work of Werner Heisenberg, but again the theory is much too complicated to be described here.

The stars come out

We can now apply the concept of turbulent motion to the discussion of what must have happened in the gaseous proto-galaxies as the result of their rotation by adopting the trend of ideas worked out by the German physicist and cosmologist Carl von Weizsäcker. From the kinetic theory of gases one finds that gaseous viscosity (or internal friction) is given by the product of gas density, thermal velocity of the gas molecules, and their free path between successive collisions. Inserting these symbols into the Reynolds formula, we arrive at the form

$$R = \left[\frac{\text{velocity of stream}}{\text{velocity of molecules}}\right] \cdot \left[\frac{\text{width of stream}}{\text{free path of molecules}}\right]$$

Velocity differences between various parts of the rotating gaseous galaxies must have been at least as large as 10 kilometers per second, whereas the thermal velocity of gas molecules at the low temperatures prevailing at that time was certainly less than 1 kilometer per second. The mean free path of molecules in the highly diluted gas forming the primordial galaxies must have been as much as 10^{16} centimeters. However, 10^{16} centimeters is only about one-hundredth of a light-year and therefore negligibly small as compared with the geometrical dimensions of the galaxies. Thus we see that the Reynolds number assumes an extremely large value, much larger than the critical value of 1000, so that *the motion of gas in the primordial proto-galaxies must necessarily have become turbulent, causing a break-up into eddies of all sizes.* Turbulent motion in gases differs from that in ordinary liquids by the fact that gaseous materials possess a high degree of compressibility. Thus the hierarchy of eddies pushing against and running into one another will result in a hierarchy of temporary local compressions of the gaseous material. These local compressions are especially pronounced when flow velocities are greater than the velocity of sound in the material in question (supersonic flow), which is exactly what is to be expected in this instance. In any gas the velocity of sound is equal to the thermal velocity of its molecules, and it has already been stated that the flow velocities in proto-galaxies (as well as those in the interstellar gas of today) were considerably larger than the thermal velocity of molecular motion.

If there were no gravitational forces, local condensations caused by turbulent motion would form and resolve without any permanent results. However, the presence of Newtonian gravity will prevent the resolution of such condensations which happen to be large enough to satisfy Jeans' criterion of gravitational instability. Instead of expanding again and mixing with other gaseous masses, such large local condensations must have continued to contract under their own weight into individual dense gas spheres. As the result of contraction the temperature of these gas spheres rose steadily and their heated surfaces began to emit first heat rays and soon afterward the shorter wavelengths of visible light. At a certain stage of contraction the central temperature of these proto-stars reached the "ignition point" of thermonuclear reactions, the powerful source of nuclear energy was switched on, and the stars settled down to the state we know today. The entire process of star formation cannot have taken longer than a few hundred million years, a small fraction of the present age of the universe. When it was finished the originally cool and dark gas masses of the proto-galaxies were transformed into the familiar swarms of shining stars.

But even though this transformation of the original gaseous proto-galaxies into the stellar galaxies of today took place billions of years ago, the galaxies still retain indications of their early youth. In fact, without the assumption that once upon a time the galaxies were made entirely of gas, there would be no explanation for their present regular shapes of rotating fluid bodies. The stars forming the galaxies of today are scattered through space so thinly that there is hardly any chance for them to influence one another's motion. It has been calculated that during the entire lifetime of the galaxy there could have been only a very few cases of two stars passing close enough to be appreciably deflected from their original tracks by the forces of mutual gravitational attraction, and there has probably not been a single case of an actual head-on collision. Under such circumstances the swarms of stars forming the galaxies could never have assumed regular ellipsoidal shapes and would have remained shapeless irregular star clouds forever. The fact that, with very few exceptions such as the two Magellanic clouds, galaxies *do* possess the regular shapes of rotating fluid bodies can be understood only on the assumption that these galactic shapes originated while the galaxies were still in the gaseous state, and that the general configuration was not changed by the condensation of the gas masses into stars. We may refer to

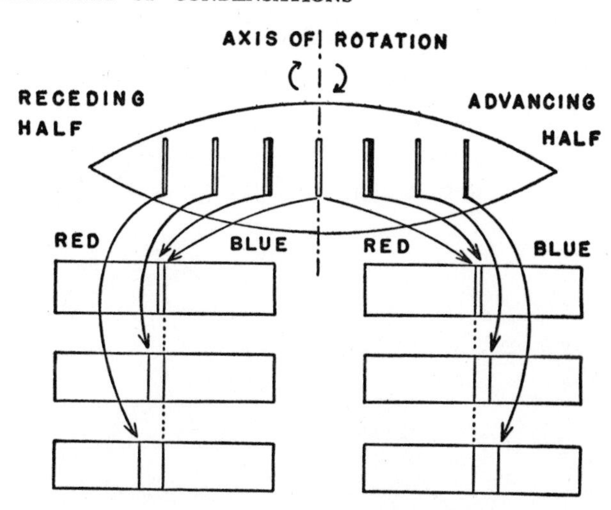

Fig. 23. The way to prove spectroscopically that elliptical galaxies rotate as rigid bodies. Red Doppler shift on left; blue on right

them as "fossilized galactic shapes," by analogy with geological fossils, such as petrified wood, which retain the exact shape and structure of a living organism although inorganic compounds were substituted for the original material a long time ago.

The theory of "fossilized shapes" has considerable interest from the point of view of the history of galactic evolution. By applying this theory, the author, working in collaboration with G. Keller and J. Beltzer, was able to explain several otherwise puzzling facts. The theory explains, for example, why the central bodies of spiral galaxies, as well as the elliptical galaxies, seem to rotate as rigid bodies, that is, with linear velocities proportional to the distances from the axes of rotation. This fact was first established observationally by measuring the Doppler shift in the light emitted from different parts of rotating galaxies, as shown in Fig. 23. The slit of a spectrograph was placed in different positions across an elliptical galaxy seen on edge, and the Doppler displacement of a spectral line measured. In all cases this displacement, which gives the linear velocity of galactic masses along the line of sight, was directly proportional to the distance from the axis. How can galaxies formed by practically non-interacting stars possess a rotation like that of a rigid body? The explanation of that paradox on the basis of the "fossilized shapes" theory is illustrated in Fig. 24. The picture at the

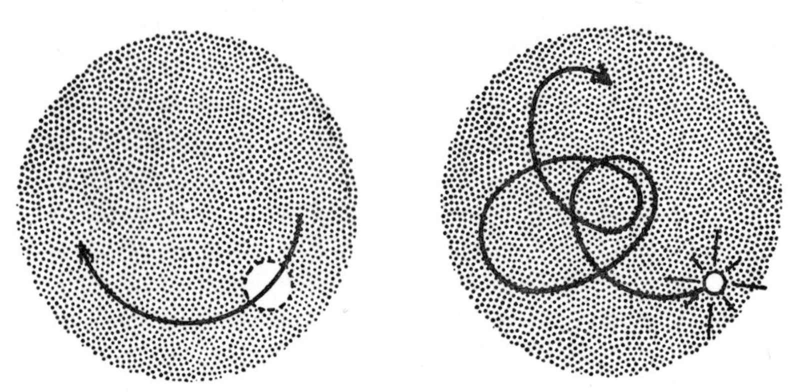

Fig. 24. What happens when a large volume of gas belonging to a rotating gas mass condenses into a star. Because it becomes much denser than the surrounding gas the star will describe the dipping elliptical trajectory shown at right

left represents the gaseous proto-galaxy with a volume of gas moving along with the rest of its gaseous body in a circular orbit around the center. When this gas volume, occupying originally several cubic light-years, condensed into a star, the pressure of the remaining gas would no longer be able to support its mass, and the new star would begin to fall toward the center with ever-increasing speed. Passing near the center at the maximum speed, it would rush out again to its original distance before making another dive. Thus the original circular motion of the dilute gas volume turns into elliptical motion of the newborn star.[4] When the gas masses of the original proto-galaxies were condensing rapidly, it must have been raining stars as blazes all through

[4] As shown in Fig. 24, the elliptical orbit of the star will move around the center too (precession). This is due to the fact that, in contrast to our solar system, the gravitational field within galaxies does not obey the inverse square law.

the galaxy. The process is actually comparable to ordinary rain caused by the condensation of atmospheric water vapor, except that rain droplets are stopped by the surface of the earth, whereas the droplets of stellar rain were destined to travel forever without hitting any solid ground. Whenever these diving stars reach their maximum elongations they are expected to possess the same tangential velocities as the original gas volumes from which they were born. Thus, by observing the light of the stars coming to their original birthplaces in the galaxy, we actually measure the "fossilized velocity" with which the original gaseous proto-galaxy was rotating. And it is logical to expect that gaseous proto-galaxies were actually rotating more or less as rigid bodies.

Old stars, new stars

As we have seen earlier, the majority of the stars were apparently formed almost simultaneously during an early stage of galactic evolution, even though some small-scale star-making is probably still going on in interstellar space. But, if the formation of stars was as easy as previous descriptions would indicate, why did not all the gas of the original proto-galaxies condense into stars, leaving the space between the stars empty and void? Why is there still a lot of gas and dust in interstellar space, in our own galaxy as well as in others?

Let's first review a few facts pertaining to the distribution of interstellar material. It is true that in the neighborhood of our sun, and indeed in the entire volume of spiral arms, there is a great deal of gas and dust, which for some reason has not condensed into stars. But what is true for the spiral arms is not true for the main central bodies of the spiral galaxies, nor does it hold true for the elliptical and spherical galaxies which do not have any arms. Recent studies, mostly made by Walter Baade of Mount Wilson and Palomar Observatories, have led to the recognition of the fact that there are *two different types of stel-*

lar population corresponding to the two different parts of the galactic structure:

1. The spiral arms (including the neighborhood of our sun) contain stars and interstellar material in about equal amounts. There are great quantities of huge gas and dust clouds, like the Great Nebula of Orion, and the space generally is so dusty that one cannot see through from one edge of the galactic disk to the other. The stellar population of these regions is characterized by the presence (even though in small numbers) of giant blue stars whose life span is so short that they must have been formed comparatively recently, much later than the main bulk of stars. This is the so-called "stellar population of Type I."

2. The central bodies of spiral galaxies, and all armless galaxies, are formed entirely of stars, with no gas and no dust present. In these regions the space between the stars is so clear that one can see through it without the slightest obscuration. The stars populating these dustless regions belong to the "stellar population of Type II," and are apparently 100 per cent original stock, with no such upstarts as Blue Giants present.

Having this information, we can answer one of our previous questions by saying that where we find a stellar population of Type II the process of star formation apparently did continue until all interstellar material was completely exhausted. But what about the gas and dust that are still present within the spiral arms? At present, it is difficult to answer this question, since we do not even know how this material originated, and for how long a time it has been there.

An interesting idea advanced by von Weizsäcker should be mentioned here because it at least provides an answer to a related question: Why doesn't the interstellar material which is still present in the spiral arms condense into stars at the same high rate as other material did during the original star-building era of billions of years ago? The answer seems to be that the formation of new stars is hampered by the existence of the stars which are already there. To prove this point we use again Jeans' formula for gravitational instability, substituting for the mean density of interstellar material the value 10^{-24} grams per cubic centimeter, and for the mass of a large star the value 10^{34} or 10^{35} grams, corresponding to 5 to 50 sun masses. From this we find that the gas temperature at which such condensations could take place must lie between 1 and 5 degrees absolute. But we know that the temperature actually maintained in interstellar space by the radiation of existing stars is in the neighborhood of 100 degrees absolute. The existing stars prevent condensation of interstellar material into new stars by heating it too much!

Another question is this: If the process responsible for the formation of the original bulk of stars does not work at present in the spiral arms, what is the origin of the few giant blue stars in these regions? There are two theories which try to account for the formation of a limited number of giant stars in the dusty regions of the spiral arms. One of these theories, proposed by Fred Hoyle and R. A. Lyttleton, is based on the so-called accretion of interstellar matter by stars moving through it. To understand this process it is best to consider the star as being at rest, and to imagine the interstellar material as flowing past it (Fig. 25). Under the action of gravitational forces streams of that material will be deflected from their initial paths and will enter the atmosphere of the star, continuously increasing its mass. It is easy to see that the higher the velocity of gas flow (in reality: the higher the velocity of stellar motion through the gas) the smaller the amount of material which can be deflected and finally captured by the star. Exact calculations show that a medium-sized star (comparable to our sun) moving with the normal stellar velocity of 10 kilometers per second will collect only a very small amount of interstellar material by this accretion process. It will be so little that it will not change the star's mass to a noticeable extent even during time intervals of billions of years. But although Hoyle's and Lyttleton's accretion

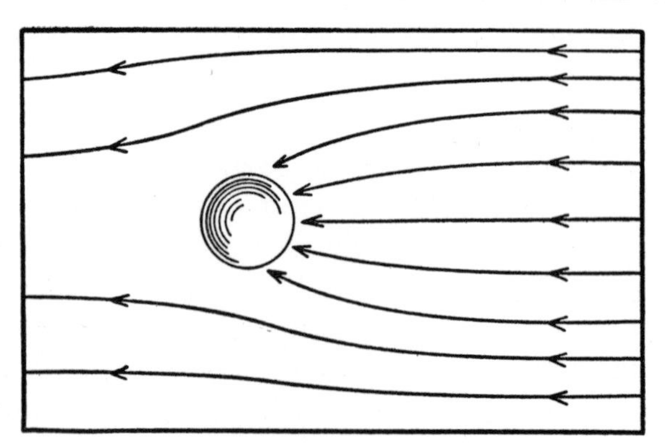

Fig. 25. Accretion of interstellar matter
(after Hoyle and Lyttleton)

process does not seem to be of much importance in the general scheme of stellar evolution it may account for some interesting developments under special circumstances. If a star passes through a comparatively dense gas-and-dust cloud, or accidentally enters a region where its velocity almost equals the velocity of gas flow (small relative velocities of star and gas), it may rapidly swell up to several times its original size. Thus it is not impossible that the short-lived Blue Giants found in spiral arms are actually old stars formed during the original process which were rejuvenated by the accretion of large amounts of additional material from some "fat" nebula through which they passed recently.

Another process by which new—really new—stars could be formed under the circumstances existing at the present time in interstellar space was suggested by Lyman Spitzer and Fred L. Whipple. Although, as we have seen, the presence of stars seems to inhibit the formation of more stars through ordinary condensation processes, the mechanism suggested by Spitzer and Whipple utilizes the radiation of existing stars. Consider a dust particle freely floating in interstellar space. It is illuminated from all sides by the stars forming the galaxy. When light falls on a surface of a material body (being reflected or absorbed) there originates a force known as "light pressure." We can visualize that force as the result of the bombardment by a multitude of light quanta which either bounce back or are stuck in the surface they hit. The pressure of light is very weak as far as bodies of normal size are concerned. Even on a brightly illuminated tennis court the light pressure experienced by flying balls does not affect their motion in the slightest, and to demonstrate the existence of that force extrasensitive equipment is required. But the smaller the body the larger the effect of light pressure, and with interstellar dust particles only a few microns in diameter this effect cannot be neglected. Since dust particles in interstellar space are illuminated about equally from all sides, the effect usually cancels out. But there will also be an effect of "mutual shadow-casting," which should be taken into consideration. If we consider two such dust particles (Fig. 26) in the field

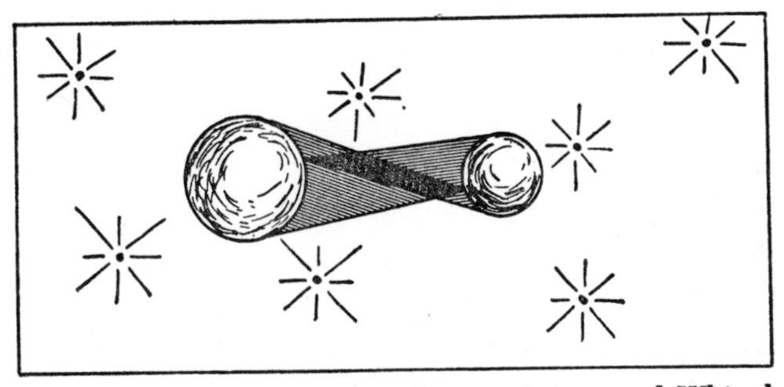

Fig. 26. Shadow casting, according to Spitzer and Whipple

of isotropic radiation coming from the surrounding stars, we find that each particle will receive fewer impacts by light quanta coming from the direction of the other one than it will from light quanta from all other directions. As the result of this mutual shadow-casting two particles will be pushed toward each other *as if* there were an attractive force between them. It is easy to show that this pseudo-attractive force will vary in inverse proportion to the square of the distance between the two particles, being similar in this sense to the force of Newtonian gravity.[5] For comparatively large particles, say a few millimeters in diameter, this "mock gravity" is very small as compared to true gravitational forces. For very small particles, however, the situation is reversed, and the "mock gravity" forces become much stronger than ordinary gravity. This is actually the case for interstellar dust particles. Consequently in interstellar space these radiation forces will drive "dust to dust," causing it to accumulate into large clouds. Once such an embryonic cloud is formed, more dust will be driven in, since all particles in its neighborhood will be in its shadow as regards the light of the stars lying on the other side. When the dust cloud becomes sufficiently large and heavy it will begin to assert its true gravitational attraction and draw interstellar gas and more dust into itself, finally developing into a nucleus of a new growing star. Detailed studies of this mechanism of star formation lead to the conclusion that it will function successfully only under a particular set of conditions which are most likely to be found inside giant intergalactic nebulae. Thus while new stars could be formed that way, it would be an exception rather than the rule.

Both the formation of new stars by the method suggested by Spitzer and Whipple and the rejuvenation of old ones according to the theory of Hoyle and Lyttleton would be expected to be unusual phenomena under present circumstances. And, in fact, observation shows that such recently born stars are exceptionally rare within our galaxy.

Clustering of galaxies and stars

As we have seen earlier in this chapter, there must have been two major differentiation processes in the history of our uni-

[5] If σ is the geometrical cross section of the particles and R the distance between them, the cone of shadow will correspond to the solid angle σ/R^2. If I is the total radiation density in space, the force acting on each particle will be

$$\tfrac{1}{3}I \cdot \sigma/4\pi R^2 \cdot \sigma = \tfrac{1}{12}\pi I \cdot \sigma^2/R^2$$

as compared with Newtonian force, which is $G \cdot m^2/R^2$

verse: the original break-up of the uniformly expanding primordial gas into billions of separate galaxies, and the condensation of the material within each galaxy into billions of individual stars. Observation shows, however, that apart from these two major events there have also been lesser and intermediate steps.

Most of the observable galaxies seem to be scattered in space more or less at random, but there are numerous cases of galaxies clustering into groups which may contain as many as several hundred individual member galaxies. One of the nearest and most thoroughly studied clusters of galaxies is the one in the constellation of Virgo. It is located only 8 million light-years away from our system, and covers most of the sky area belonging to the constellations of Virgo, Coma Berenice, and Leo. At this distance the recession velocity due to the expansion of the universe amounts to only 700 miles per second, and since the random velocities of individual galaxies forming that cluster are sometimes as high as 1500 miles per second, many of its members move toward us, showing a blue shift of the spectral lines. This is one of the cases already mentioned, where random speeds of individual members exceed the regular recession velocity due to expansion. In fact, the Virgo cluster is so close to us that the question has been raised whether our own system of the Milky Way should not be considered as one of its members. Whether or not we belong to that giant family of galaxies, there is no doubt that the system of the Milky Way is not a lone wolf in space. In fact, it is a member of the so-called "local group" which consists of three spirals (the Andromeda Nebula being one of them), six elliptic galaxies, and four irregular or shapeless galaxies (including the Large and the Small Magellanic Clouds).

We know only a couple of dozen other clusters of galaxies that contain as many members as the Virgo cluster (one of them, located in the constellation of Corona Borealis is shown in Plate VI) but there are over a hundred lesser groups similar to our "local group," and virtually thousands of still smaller associations, sometimes limited to triplets or pairs of galaxies. What is the physical reason for such grouping?

There are two possible ways in which galactic clustering may have originated. We may assume, as von Weizsäcker does, that the primordial gas in the original expanding universe was not as homogeneous as we assumed in earlier discussions. One could indeed imagine that the process of regular expansion was overlapped by some kind of turbulence, and that the regular motion of masses of gas was broken up into a large number of turbulent eddies of different sizes. The standard size of a galaxy would correspond, according to such an assumption, to the smallest size of an eddy which can be held together by its own gravity, and the process of galactic formation would be similar to the process which led to the formation of stars within individual galaxies. The clusters of galaxies should then be considered as resulting from larger eddies which existed within the primordial gaseous material. The only unsatisfactory part of such an explanation is that we have to postulate the existence of turbulence in the primordial material instead of deriving it as a natural consequence of the expansion process. Such a postulate may well turn out to be the only way to account for the presence of turbulent motion in the later stages of expansion.[6] Should this be the case we will have to relax our requirements concerning the simplicity of the initial assumptions.

Another possible explanation of the observed clustering of galaxies is the assumption that at the moment of their origin they were scattered through space in random fashion, and that they started grouping into clusters at a later date under the action of mutual gravity. The mathematical problem relating to

[6] In the previous discussion we tried to attribute the turbulence within proto-galaxies to the rotation caused by the irregularities of the breaking-up process in a completely homogeneous expanding material. But it is not quite certain that this assumption would stand rigorous mathematical analysis.

the temporal behavior of a multitude of gravitating points originally scattered at random over an infinite space is extremely complicated. It was recently tackled by S. Ulam, who was able to show that, in the case of a simplified one-dimensional model (with the gravitating points distributed along a straight line), one should expect the clustering of points into groups of various sizes. However, the extension of Ulam's result into two- or three-dimensional space seems to present insurmountable mathematical difficulties.

Intermediate clustering is also shown by the stars within a galaxy. We have already mentioned the "galactic clusters" [7] of stars (such as the Taurus cluster shown in Fig. 2), which contain many stars with the same proper motion through space. It seems very probable that all the members of this type of group have condensed from one giant gas-and-dust cloud (one of the large-scale turbulent eddies existing within our galaxy) and that they are all moving with the same velocity that the original cloud had. Due to perturbations caused by other stars and the effect of differential rotation in the region of spiral arms, such clouds, and the resulting star clusters, were never able to assume any regular shape, and are being gradually dissolved as the galaxy grows older.

There are, however, other star clusters which are located in the quiet regions of space away from the main stellar traffic of the galactic plane. Being undisturbed by other stars, they assume regular spherical shapes and are known as "globular clusters." One such cluster, found in the constellation of Hercules, is shown in Plate VII. In such clusters the distances between individual stars are comparatively small, so that the mutual gravitational forces have a good chance to change their motion, and to arrange the entire system into a beautiful spherical pat-

[7] There is an unfortunate confusion of terminology between clusters of galaxies and galactic clusters. The latter are the clusters of stars within the spiral arms of the galaxies.

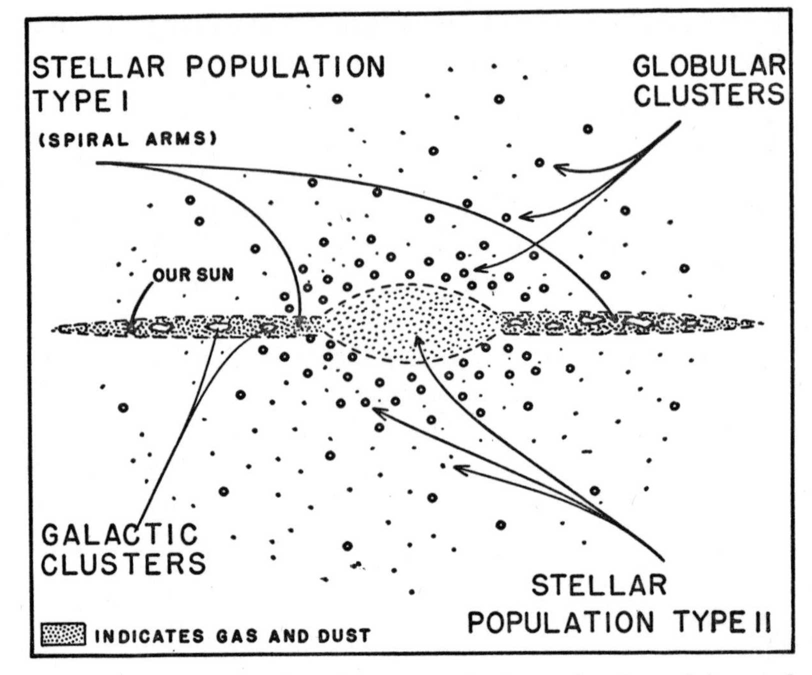

Fig. 27. Cross section of a galaxy perpendicular to the plane of the spiral arms

tern identical with the shape that would be assumed by a large mass of gas freely floating in empty space.

Figure 27 shows a general scheme of our galaxy indicating the location of galactic and globular clusters in respect to the regions characterized by the two types of stellar population previously described.

Planetary systems

We now come to a consideration of the evolution and properties of individual stars, which are, after all, the most important single units of our universe. Of all the stars in the universe, the sun is to us the most important one, because it is nearest to us. An equally important fact about that star is that it possesses a system of planets. In fact, only a few centuries ago, the sun with its plan-

Plate VI. Cluster of faint galaxies in Corona Borealis. Galaxies can be recognized by their spindle shape. The round dots are foreground stars

Plate VII. Globular star cluster in Hercules

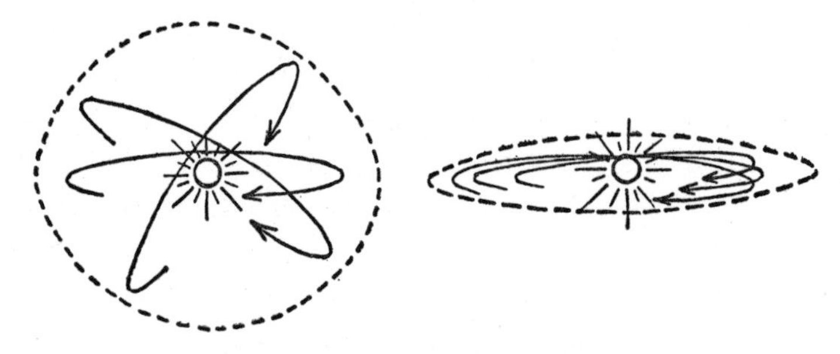

Fig. 28. How material originally moving in different planes flattens into a disk because of mutual collisions among the particles

etary system was about all that astronomy could study in detail; consequently the early cosmogonic theories were limited to the problem of the origin of the solar family. Scientific cosmogony starts historically with the views first expressed by the French naturalist Georges de Buffon, who visualized the birth of planets as the result of a glancing collision of our sun and a passing comet. A rather different and more elaborate hypothesis was expressed somewhat later by the German philosopher Immanuel Kant, and by the French mathematician Pierre de Laplace. According to their hypotheses, the birth of planets was not the result of an accidental hit-and-run encounter, but rather a part of a normal process to be expected in the life of almost every star. Both Kant and Laplace assumed that the young sun was surrounded by a thin lens-shaped gaseous envelope (solar nebula) which later condensed into individual planets as we know them today. Such an assumption is in perfect agreement with modern views concerning the formation of stars from diffuse interstellar material. Indeed it is logical to expect that some portions of the material forming the turbulent eddy which later became our sun would be prevented by their large angular velocities from falling into the center. An almost spherical rotating envelope of the kind that must first have surrounded a newborn sun would rapidly flatten out because of the collisions between its parts moving in different planes (Fig. 28). Due to gravitational attraction between different parts of that diffuse disk, its material might then condense into separate planets.

The Kant-Laplace hypothesis, which dominated scientific thought for over a century, was, however, severely criticized by the British physicist James Clerk Maxwell, who thought he had proved that such a condensation could never have taken place. Gravitational condensation as visualized by Kant and Laplace was a special case of a more modern notion, namely Jeans' "gravitational instability," which we have already discussed several times in this book. We have seen that, for any given temperature and density of a gas, there is always a minimum size for condensations which could hold together by mutual gravitational attraction. But in the rotating gaseous disk that allegedly sur-

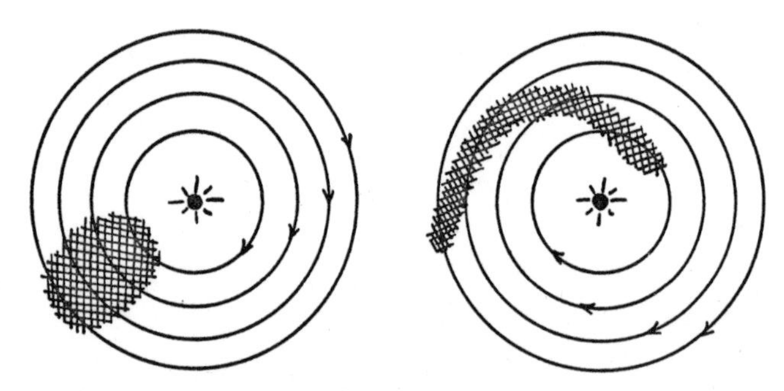

Fig. 29. How a beginning condensation may break up because of differences in the angular velocities of its parts

rounded the young sun the situation was complicated by the fact that in accordance with Kepler's laws different parts of the disk had to rotate with different angular velocities. Thus, if the rudimentary condensation was too large, its different parts might have been pulled away from each other by the shearing forces of differential rotation, and it would have dissolved again. Such a shearing process is shown in Fig. 29. The relative velocities of differential rotation do not depend on the density of the rotating disk, but the gravitational forces trying to keep it together are proportional to this density. Consequently there has to be a certain critical value for the density above which rudimentary condensations could hold together in spite of the disruptive forces of rotation.[8] Assuming that the mean density of the primordial disk was equal to that obtained by spreading the combined mass of all planets (about 0.001 sun mass) uniformly over the plane of the ecliptic (10^{-11} grams per cubic centimeter), Maxwell found that the forces of differential rotation would break up any condensation as soon as it began to form. He was able to show that no gravitational condensations could ever have appeared unless the amount of material in the disk was at least one hundred times the total mass of all planets.

This apparent contradiction forced the cosmogonists to abandon the views of Kant and Laplace and to return to the original collision hypothesis of de Buffon, substituting a passing star for a comet. However, this rejuvenated collision hypothesis, developed simultaneously by Sir James Jeans in England and by Forest Ray Moulton and Thomas C. Chamberlin in the United States, ran into still more serious difficulties, and was never able to make much headway. The paradox was finally resolved toward the end of World War II by von Weizsäcker, who indicated that Maxwell's old objection is no longer valid in view of our improved knowledge of the chemical constitution of cosmic matter. For lack of information to the contrary it was customary to assume that the sun, the other stars, and the interstellar material consisted mostly of iron, silicon, and other "terrestrial elements," just like the earth. These views are now completely changed (see Chapter III) and we know that the "terrestrial elements" form only about 1 per cent of all matter, the rest being essentially a mixture of hydrogen and helium. The material of which the planets were built represents only about one-hundredth of the total original material of the disk, raising its former mass from

[8] Mathematically this critical density is given by

$$\rho_{crit} = w^2/G$$

where w is the angular velocity of rotation, and G the Newtonian constant of gravity.

one-thousandth of the solar mass to one-tenth. This brings the mean density of the original disk to exactly the value which, according to Maxwell, would permit gravitational condensations in spite of the forces of differential rotation.

The difference between the old Kant-Laplace theory and the modern version proposed by von Weizsäcker lies in the recognition of the essentially different behavior of the gaseous (hydrogen-helium) part of the disk, and the dust portion of small solid particles of "terrestrial materials" identical with those which we now find in interstellar dust clouds (Fig. 30). Because

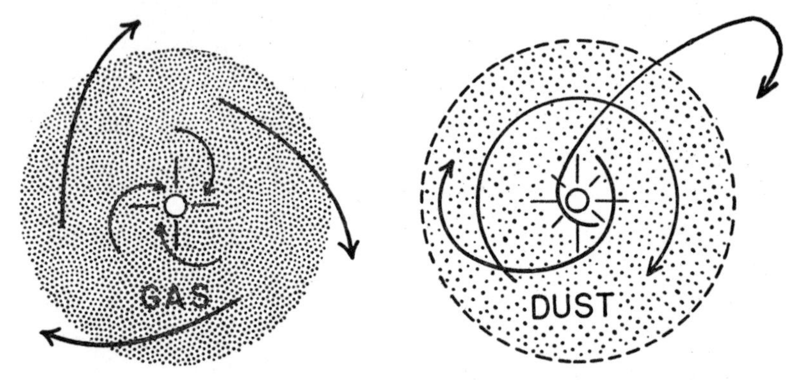

Fig. 30. The behavior of gas and dust particles in the original solar nebula, according to von Weizsäcker

of the viscosity of gas, the original gaseous envelope would have had a tendency to rotate as a rigid body, that is, with linear velocities proportional to its distance from the sun. This would communicate such high velocities to its outer parts that they could no longer be held by the forces of solar gravity and would be thrown out into space. On the other hand, the inner part of the disk would be slowed down, and its material would fall into the sun following spiral trajectories. Consequently the original gaseous envelope must have gradually dissipated. It can be calculated that its mass was cut in half every 5 million years. This reduced its density from the original value of 10^{-9} grams per cubic centimeter to the much smaller value of 10^{-22} grams per cubic centimeter in about $2 \cdot 10^8$ years. We can see that the gaseous envelope of the sun, although it played an important role in the process of planetary formation, must have been gone completely long before our time.

The fate of the dust particles in the original disk, however, was entirely different from that of the gas molecules. Being small and dense, these particles must have been moving more like the planets of today, describing various elliptical orbits around the center of the sun. However, since there must have been a tremendous number of such particles (10^{30}, if each weighed a few grams) the traffic must have been extremely heavy and collisions between particles very frequent. When two particles of about equal size collide with meteoric velocities, both may be expected to be completely pulverized, or rather turned into vapor which will later condense again into microscopic dust grains. But when a small particle strikes a large one it sticks to it, thus adding to its mass. The result of these processes must have been the formation of larger and larger chunks of solid material which kept sweeping the space around them and absorbing the smaller pieces. While the original aggregations of dust were caused exclusively by direct head-on collisions between the particles, the chunks above a certain fairly large size must have begun to attract smaller particles by true gravity, thus collecting material from a rather wide area. The situation resembles the growth of large industrial monopolies which swallow up all the smaller companies, but since there is no antitrust law in the cosmos, the process continued until only a few "big fellows" remained—too far apart to interfere with one another.

The process of dust aggregation must have been running concurrently with the dissipation of the original gaseous disk, and, in fact, the two have comparable time scales. It must have taken just a few years for the original microscopic dust particles to grow to pebble size (1 centimeter in diameter) and about 10^8 years to build a large planet like Jupiter. The mass of individual planets must have been determined by the total amount of material present in the regions where they were formed, so that we should expect that the biggest planets would develop about halfway between the sun and the outer rim of the disk; close to the sun the density may be high but the available volume of space is much too small, and at the rim, while there is plenty of space, the density is much too low. This idea agrees nicely with the observed fact that the inner and outer planets (Mercury, Venus, Earth, Mars, and Pluto) are much smaller than the intermediate ones (Jupiter, Saturn, Uranus, and Neptune).

Turning to the details of the formation of planets, we get into rather deep water at once, since the combined motion of gaseous masses and growing dust particles must necessarily have been complicated. The great merit of von Weizsäcker's theory lies in its recognition of the important role played by turbulent motion. The conditions for laminar flow were far from being satisfied in the material of the disk, so that the motion must have been broken up into a multitude of eddies. The situation was similar to that which led to the formation of stars within the primordial gaseous proto-galaxies. But there was an important difference. The eddies in the proto-galaxies measured only a fraction of 1 per cent of the total galactic thickness. In the solar nebula, however, the smallest gravitationally stable eddies must have been comparable in size to the entire thickness of the gaseous disk. Thus the original galactic clouds broke up into billions of stars, whereas the material of the solar disk gave birth to less than a dozen planets.

Von Weizsäcker and, after him, Ter-Haar, Chandrasekhar, and Kuiper made considerable advances in understanding turbulent motion within the solar nebula. Their results can be described here in only a very general form. The key to understanding the complex motion that took place lies in singling out a group of particles with about the same rotation periods (which also means the same mean distance from the sun) and considering their motion in a coordinate system which rotates around the center of the sun with the same period. As seen from such a rotating coordinate system, its particles, moving with constant speed along their circular orbits, will be at rest. But particles moving along elongated elliptical orbits run ahead when they are close to the sun, and lag when they are far from it. As seen from the coordinate system, these particles describe certain closed curves, which are larger in size for the more elongated orbits. The situation, resembling the famous "system of epicycles" of ancient astronomy, is shown in Fig. 31. In the course of time the motion of the multitude of dust particles tends to a state characterized by the minimum number of collisions. Such a state can be represented in our rotating coordinate system by a pattern of non-intersecting epicycles, some of them lying inside one another. Each group of particles belonging to a concentric system of epicycles represents a whirlpool in the medium of the solar nebula, or, in other words, a turbulent eddy. Since we have chosen only the particles possessing the same periods

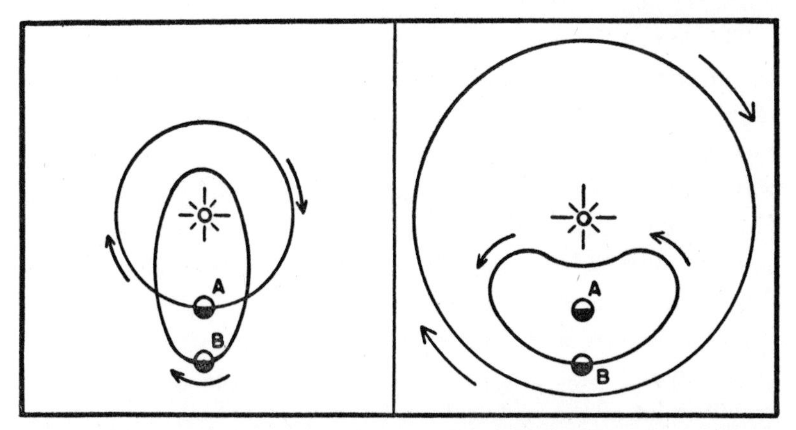

Fig. 31. How a circular motion (A) and an elliptical motion (B) of a dust particle look from the point of view of a coordinate system rotating with the same period

and the same mean distance from the sun, this pattern looks somewhat like a necklace made of flat sea shells. There are many such circular necklaces, placed one inside the other, and rotating with different periods; shorter periods close to the sun, and long periods far from it. Figure 32 shows von Weizsäcker's original picture (top) and the much more complicated system, constructed by Kuiper, with random eddy sizes distributed according to Kolmogoroff's law of turbulence (bottom). Only the largest eddies will be kept together by the force of Newtonian gravity, and thus escape subsequent dissolution. Within these eddies the process of dust aggregation will continue at a rapid rate, resulting finally in the growth of planets. Studying in detail the properties of the original solar nebula, and the turbulent motion produced by its rotation, it was possible to obtain the correct sizes of the different planets of the solar system. It was also possible to give a reasonable explanation of the famous Bode-Titius rule of planetary distances, which states that within the planetary family (considering the asteroid ring between Mars and Jupiter as the remainder of an old planet) *the distance of each member from the sun is approximately twice the distance of the previous one.*

This is not the place to go into further details of the theory; therefore we will only indicate a few of its most interesting consequences in passing. First, as was mentioned in Chapter III, such a process of formation results in different chemical constitutions for the smaller and larger planets. While "rim" planets (Mercury to Mars on the inner side, and Pluto on the outer side) never grew massive enough to attract much interstellar gas, and thus remained essentially rocky structures, the "halfway" planets (Jupiter, Saturn, and, to a lesser degree, Uranus and Neptune) grew beyond that limiting size and were able to capture by gravity a lot of the material from the original gaseous shell before it was dissipated into the surrounding space. (See Plate III, in which the internal structure of Jupiter is indicated.)

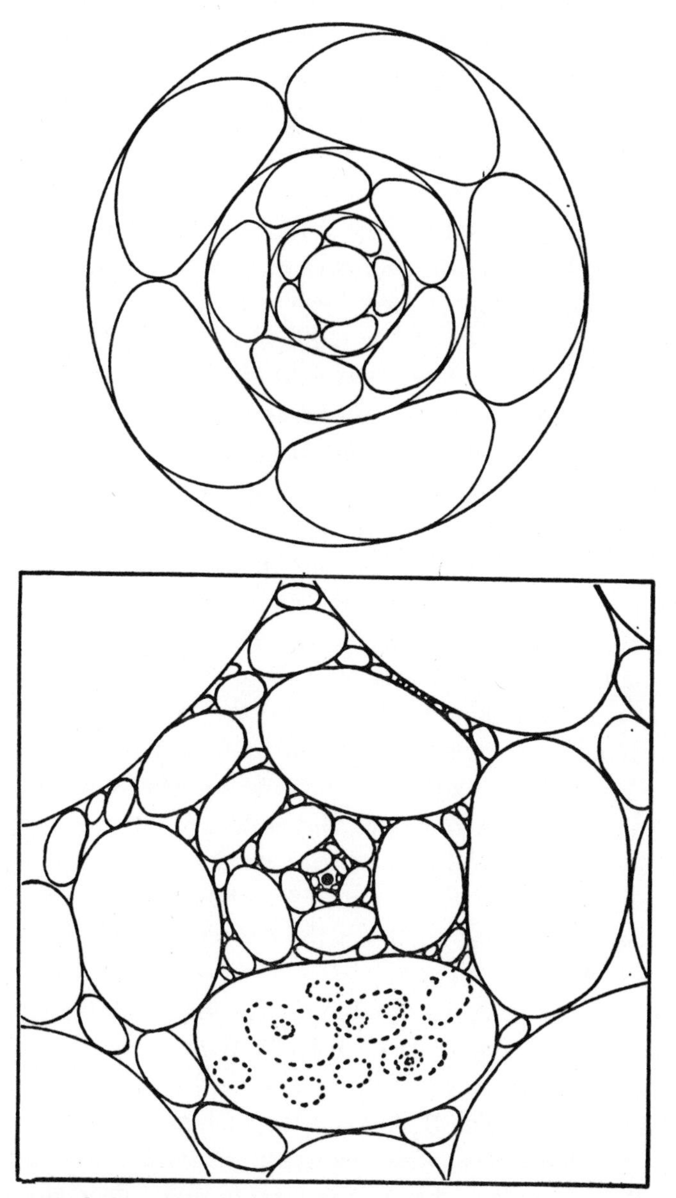

Fig. 32. Turbulent eddies in the solar nebula, (top) according to von Weizsäcker's original theory, (bottom) according to Kuiper

The planet which must once upon a time have existed between Mars and Jupiter was presumably broken up some time in the past (probably by tidal forces of Jupiter), and its fragments are now moving in the neighborhood of the old orbit where they form the ring of asteroids. Some pieces belonging to the asteroid ring venture far, far away from its domain, occasionally falling on the surface of the earth in the form of meteorites. The chemical constitution of meteorites strongly suggests that their material must have been solidified under very high pressures such as would be found in the interior of a planet.

No such obvious relationship exists with the other group of members of the solar system: the comets. Though they are about the most spectacular objects which can be seen in the sky by the naked eye, comets constitute a very insignificant part of the solar system. They were probably formed from the outermost fringes of the original solar nebula, and are mostly composed of chemical compounds of light elements, such as water, ammonia, and various hydrocarbons. Probably some 20 billion comets are moving (mostly far beyond the orbit of Pluto) within a sphere about 3 light-years in diameter. But, since the mass of an average comet is only 10^{16} grams, they all add up to only about one-tenth of the mass of the earth. Occasionally comets from this large reservoir come close to the sun, and under the action of its radiation develop beautiful tails which spread horror among superstitious natives of Africa and inspire such fantastic stories about misbehaving celestial bodies as the one by Immanuel Velikovsky.

Last but not least, we should mention that, with the possible exception of our moon (see Chapter I), the systems of planetary satellites were developed by a process very similar, if not identical, to that by which the planets themselves were formed.

In accordance with these views, in contrast to the old collision theory, we should expect any star to have a good chance to procure for itself a planetary system. And, indeed, observation seems to indicate that this may well be the case. The detailed studies of the proper motion of nearby stars, such as Barnard's star (6.1 light-years distant) and 61 Cygni (11 light-years distant), carried out by P. van de Kamp at Sproul Observatory and K. A. G. Strand at Northwestern University, show that their motion through space is not a perfectly straight line. This fact indicates the presence of an invisible companion, but the observed deviations from the straight line are so small that the mass of the companion cannot be much larger than that of Jupiter. The order of magnitude is of planetary rather than of stellar scale. It is also quite possible that these stars possess a number of smaller planets similar to our earth, but the precision of measurement presently obtainable is not high enough to detect them.

CHAPTER V

The Private Lives of the Stars

Stellar symbiosis

In Chapter IV we discussed the sequence of successive condensations all the way from the original proto-galaxies to individual stars, planets, and the satellites of those planets. A fact not stressed so far is that single stars like our sun—with or without planetary systems—are an exception rather than the rule. About 80 per cent of all stars are multiple stars—chiefly double stars (binaries), but in some cases even triple or quadruple stars.[1] Consequently the problem of multiple stars assumes primary importance in any theory of stellar origin and evolution.

Astronomers classify double stars as "visual," "spectroscopic," and "eclipsing" binaries, but this classification refers to the method of observation rather than to the physical nature of the pairs. If two stars are sufficiently far apart, as compared with their distance from us, to be seen as two stars through the telescope, the pair is called a visual binary. In such a case we can study their orbital motions in detail and obtain a maximum of information concerning the mechanical properties of the system. However, if the two components of a binary are very close together, or if the binary system is very far away, telescopic observation will show only one luminous point, and the binary nature of the star can be detected only by means of the spectro-

[1] Of the catalogued stars only about 20 per cent are listed as multiple, but this is probably due to observational selection.

scope. Component A is moving toward us during half of its rotation period and away from us during the other half; and component B is moving away from us while A moves toward us, and vice versa. Therefore every line in the spectrum will show a periodic Doppler-effect splitting. This splitting provides us directly with the period of rotation of the components and with their orbital velocities, or, more precisely, the geometrical projection of the orbital movement on the lines of sight. Thus the spectroscope, in the case of these spectroscopic binaries, not only reveals the fact that the star is a binary but also provides about the same information that is obtained by direct observation for visual binaries.

In exceptional cases a double star may happen to be oriented in space in such a way that the line of sight from the earth very nearly coincides with the plane of the orbital motion of the components. In this case the two stars will periodically eclipse each other, providing additional information about the dimensions of their disks and often about the nature of their atmospheres.

From the point of view of stellar evolution, binaries are of special interest only in those cases where the distance between the two companions is small enough—it must be comparable to their diameters—to cause strong physical interactions between their atmospheres. A typical example of such a close, or "contact," binary is furnished by the second brightest star in the constellation of Lyra, Beta Lyrae, which is composed of a giant blue star and a much smaller yellowish companion. The separation of the two is roughly equal to the diameter of the main component. Figure 33, which is based on the observational results obtained by Otto Struve and the theoretical studies by Gerard P. Kuiper, shows what happens to the outer layers of these stars as the result of Newtonian gravity. It seems that a massive stream of hot gases leaves the surface of the larger star and moves toward the smaller companion with an average velocity of about 300 kilometers per second (180 miles per second).

Fig. 33. Ejection of gaseous material by Beta Lyrae, according to O. Struve and G. P. Kuiper

But because of the rotation of the entire system this stream misses its target and passes behind the smaller star, which, however, strongly deflects the stream by its gravitational field. A portion of the stream is bent all the way around and returns to its original source. But other, faster-moving portions of the stream are thrown clear out into space, moving along a gradually expanding spiral path. It is probable that some of this material is captured by the smaller component, leading to a gradual increase of its mass.

In the present state of our knowledge it is difficult to tell what the evolutionary future of this and similar systems will be. In particular, observation has not yet revealed whether this exchange of material between the two stars of a close binary will lead to an increase or to a reduction of the distance between them. If the distance increases, we may expect that the gas stream will gradually lose volume and finally peter out completely. If the distance decreases, an actual contact must result, causing the smaller star to fuse with the larger into a single body.

Important as the origin of multiple systems is to general cosmogony, we are still unfortunately without a satisfactory theory to account for it, although a large amount of work has been expended on the problem. It seems very probable that the phenomenon is somehow connected with the presence of high angular velocities in the original material from which the stars were formed. It is a fact that the percentage of binaries is much higher among the stars belonging to the spiral arms (Stellar Population I) than among the stars populating the almost spherical region surrounding the central galactic body (Stellar Population II).[2]

It would seem logical to think that binaries must have originated by way of fission of single rapidly rotating stars. In fact, when a newly formed star contracts to a smaller and smaller radius, its rotational velocity gradually increases and is bound to reach a critical value at which the centrifugal force becomes strong enough to tear the star apart, thus forming a close-contact binary. The further evolutionary history of the binary will then be determined by the gravitational interaction between the two components, which may result in a steady increase of the distance between them. As long as the two stars are very close together this interaction will take a disruptive form similar to that described for Beta Lyrae. When the distance has grown larger there will be ordinary tidal forces similar to those which are driving the moon away from the earth (see Chapter I).

The main objection to the fission theory of the origin of binaries lies in the fact that such a break-up cannot be expected to take place in stars with normal distribution of densities in their bodies. It was shown mathematically by Sir James Jeans that fis-

[2] As Fig. 27 shows, stars of Stellar Population II are scattered more or less uniformly in all directions around the central body and are moving mostly radially (toward or away from the center), showing little angular rotation and small differential velocities.

sion of a rapidly rotating body can take place only if the body has approximately uniform density from center to surface. If there is a decided concentration of matter in the central region, the break-up process will take an entirely different course. The star will then develop a kind of sharp-edged equatorial bulge and the material will be ejected from the edge in the form of a fine spray. The difference between the two cases is illustrated in Fig. 34. We know that all normal stars do possess highly

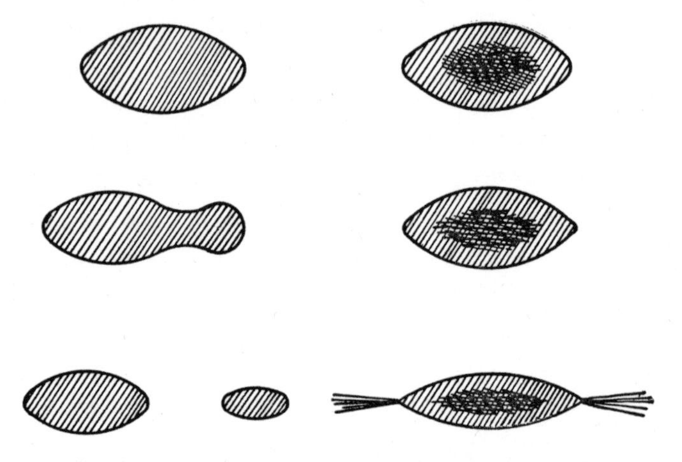

Fig. 34. What happens to a rotating star, (left) with a uniform density, (right) with a central condensation

condensed centers; in our sun the density in the central region exceeds the mean density by a factor of almost 100. We can also observe many cases of rapidly rotating stars which actually eject material from their equatorial bulges. It would seem, then, that the fission process is not an acceptable explanation.

However, all our information pertains to stars which were formed some time ago and have had enough time to adjust the distribution of matter in their interiors. We know nothing about the density distribution in a star *in statu nascendi,* immediately after its condensation from the original diffuse gaseous material. Such a "proto-star," while it had contracted sufficiently to have

Plate VIII. "Ring nebula" in Lyra, a typical planetary nebula

Plate IX. An expanding nebula resulting from the explosion of Nova Persei

a high velocity of rotation, might still possess a rather uniform distribution of mass throughout its interior. Under such conditions the fission process might still take place, producing two components which would later readjust to normal density distributions within their bodies.

The trouble with the problem of the origin of binary stars, as well as with many other problems of cosmogony, is that we never have a chance to observe the formation process itself. We are face to face with the finished product. We are thus deprived of many important pieces of evidence which could be obtained by direct observation of the formation. And the mathematical theory of the processes which can be expected to take place in a gravitating turbulent medium is tremendously complicated and much of it is out of reach of ordinary analytical methods. It can be hoped, however, that much progress will be made in the near future by the use of modern electronic computing machines with which one can tackle complex hydrodynamical problems with as much ease as a gifted schoolboy can take on a routine problem in algebra. Such studies, when completed, should also tell us why the condensation of the original diffuse material led in some cases to double or multiple stars, while in others a single star with a planetary system was the result.

Sources of nuclear energy

When the stars were first formed by the condensation of the original proto-galaxies, they were only gigantic, slowly shrinking spheres of lukewarm gaseous material. However, as a result of progressive condensation and the simultaneous liberation of vast amounts of gravitational energy, these gaseous spheres rapidly heated up and their surfaces began to emit visible light. Each star was running the course of its "contractive evolution" from a rather dilute red-hot stage with comparatively low luminosity toward the highly condensed and brilliantly luminous white-hot stage. In describing the observable properties of stars it is customary to use the so-called Hertzsprung-Russell diagram (Fig. 35) in which the luminosities of stars (or rather the logarithms of those luminosities) are plotted against the colors of the stars as characterized by their surface temperatures. In the diagram the tracks of contractive evolution are represented by straight lines running from the lower right-hand to the upper left-hand corner.

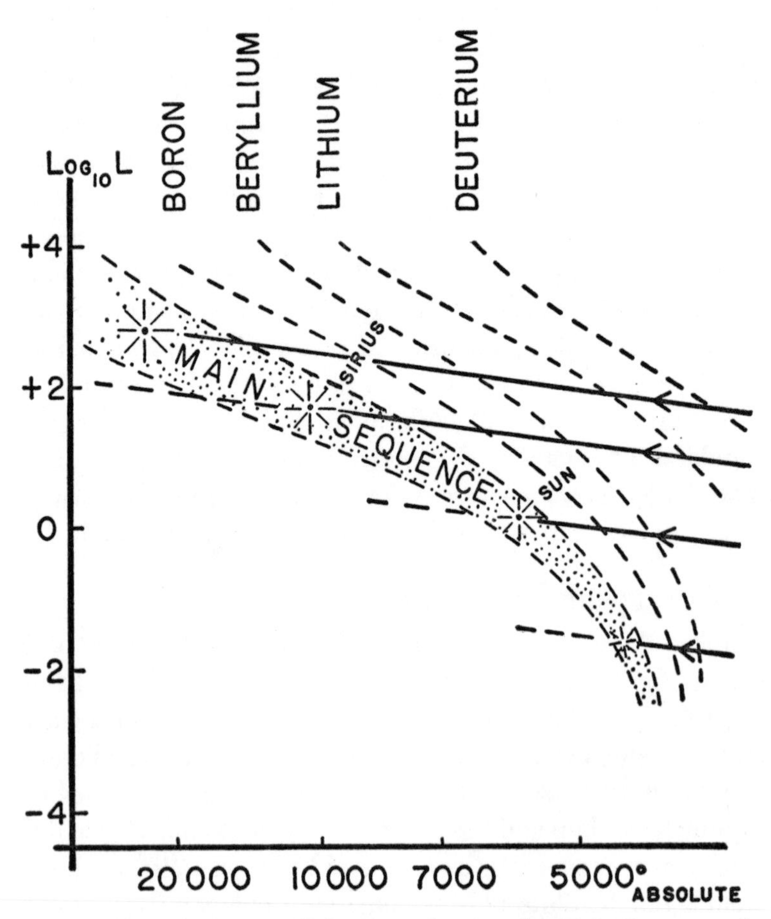

Fig. 35. Hertzsprung-Russell diagram, showing the Main Sequence, the tracks of contractive evolution, and the "stopover lines" due to the reaction with lighter elements

When a star evolves by gravitational contraction its surface temperature and its luminosity increase continuously and rather rapidly with time. If, for example, the present state of the sun were just a transitory stage of such evolution by contraction its luminosity could be expected to double every 10 million years. It follows that 10 million years ago the amount of solar radiation striking the surface of the earth would have been insufficient to maintain surface temperatures above the freezing point of water.[3] If this were true, the oceans of early geological epochs would have been merely solid chunks of ice, which melted only comparatively recently.

We know, however, that this cannot be true. Paleontological evidence shows that the temperature of our planet has remained very close to the present value for a period of at least 1 billion years, the time required for the development of the present forms of life. We are forced to conclude that our sun, and presumably all other stars, possess some other and richer source of energy which enables them to maintain a status quo for very long periods of time. And there can hardly be any doubt that this other source of energy has to be sought in the nuclear transformations which take place in the hot interiors of the stars. As soon as the central temperature of a rapidly evolving contracting star reaches the "ignition point" of nuclear reactions, contraction stops and the star remains in its new nuclear status quo until this source of energy is finally exhausted.

From the Hertzsprung-Russell diagram we can see that most of the existing stars are located along a narrow ribbon known as the Main Sequence, which runs across the tracks of contrac-

[3] According to the Stephan-Boltzmann law the surface temperature of the earth (absolute scale) varies as the fourth root of solar luminosity. Since the present temperature is of the order of 300 degrees absolute, the temperature corresponding to one-half of the present solar radiation would be

$$300 : \sqrt[4]{2} = 300 : 1.2 = 250 \text{ degrees absolute}$$

which is well below the freezing point of water.

tive evolution. This, then, must be the sequence of the points at which sources of nuclear energy are switched on for stars of different masses. In order to ascertain which particular nuclear reaction is responsible for stopping contraction, we must know what physical conditions exist in the central regions of the stars of the Main Sequence. This information is provided by the theory of stellar structure developed by the British astrophysicist Sir Arthur Eddington and his successors. At first glance the problem of finding the temperatures and pressures prevailing in the interiors of stars appears very difficult indeed. Actually, however, we can get much more extensive and reliable information about conditions inside a star several hundred light-years away than we have about the interior of the earth only a few thousand miles under our feet. This is because the earth is made up of solids and molten matter, while stars consist entirely of gases which obey much simpler physical laws.

The gaseous material which forms the interior of a star is even simpler in its physical properties than is atmospheric air. Because of the extremely high temperatures, the atoms of that gas are dissociated into stripped atomic nuclei and free electrons. With our present knowledge of atomic physics we can predict the characteristics of such a gas with a high degree of certainty and can derive reliable and simple formulae for its mechanical, optical, and electrical properties. Possessing these formulae and starting with physical conditions observed at the surface of a star, we can determine step by step the conditions prevailing in deeper layers, ending with figures for temperature, pressure, and density at the very core. These calculations were first made for our own sun and showed that at the sun's center the temperature must be around 20 million degrees and the density about one hundred times that of water.[4] Applying the same method to

[4] The reader should not be surprised that material of such high density is still considered a gas. The characteristics typical of the gaseous state are present whenever the distance between particles is greater than the

the stars of the Main Sequence, Eddington found that, no matter whether we select a very faint or a very bright star, the central temperature is always close to 20 million degrees. Of course there is a difference between faint stars and bright ones, but it is surprisingly slight. Obviously this value of 20 million degrees represents the ignition temperature of thermonuclear reactions which maintain the constant luminosities of the stars of the Main Sequence.

In order to establish the particular reaction involved and the species of chemical elements participating in it we have to turn to the results of nuclear physics, both experimental and theoretical.

It is known that nuclear reactions take place whenever two nuclei collide with a velocity sufficient to penetrate the barrier of electric repulsion caused by their positive charges. In the various types of "atom smashers" used in physical laboratories, these high-velocity collisions are produced by accelerating particles in high-voltage fields. In the stars the same kind of collision results from the intensive thermal motion of particles caused by extremely high temperatures. Using the quantum theory of nuclear reactions developed by E. U. Condon and R. Gurney, and independently by the author more than 20 years ago, one can calculate the rate of energy liberation for different elements at different densities and temperatures. Such calculations were carried out for the first time in 1929 by R. Atkinson and F. Houtermans. They indicated that the only nuclear reactions which can produce the observed amounts of energy under the conditions of the sun's interior are the reactions between hydrogen and the nuclei of light elements.

At that time, however, the empirical knowledge of different types of nuclear reactions was in its infant stage; and it was

---

size of the particles. Since the atoms of matter in the interior of a star are broken up into much smaller particles (electrons and nuclei), the material remains gaseous at much higher densities.

more than 10 years before the "solar reaction" could be established in complete detail.

We now know that the process which supplies the sun with nuclear energy consists of a sequence of reactions which has been called the "carbon cycle." This chain of nuclear reactions,

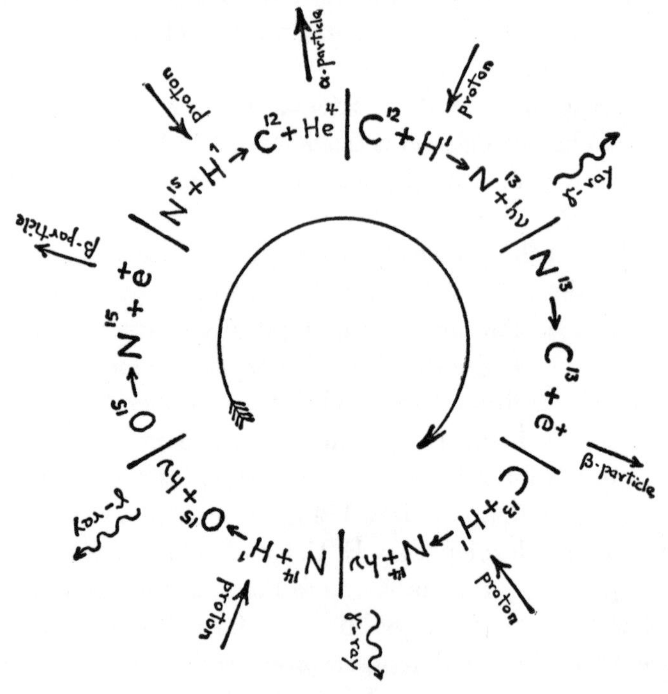

Fig. 36. The cyclic nuclear reaction responsible for the generation of energy on the sun

which was suggested independently by H. Bethe and by Carl von Weizsäcker in 1938, is shown in Fig. 36. Its net result is the transformation of hydrogen into helium through the catalytic action of carbon and nitrogen.

A "competing" reaction, first proposed at about the same time by C. Critchfield (see Fig. 37), achieves the same transformation, namely, hydrogen into helium, without the aid of any "nuclear catalyzer." In our sun, about 85 per cent of the liberation of energy is produced by the carbon cycle, while Critchfield's hydrogen-helium (H–H) process takes care of the remaining 15 per cent. Since the two reactions depend to a large extent upon temperature, their relative importance varies for different stars, depending on mass and luminosity. In stars which are

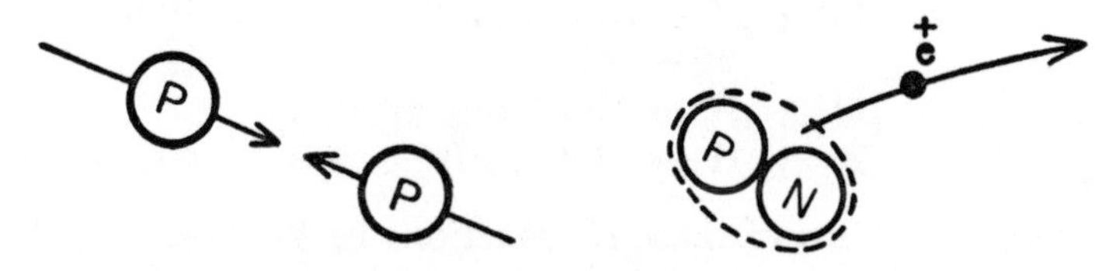

Fig. 37. Critchfield's H-H process

much brighter than our sun, as for example Sirius, the carbon cycle is responsible for virtually all the energy production, while stars much fainter than our sun rely almost exclusively on the H–H process.

Both these reactions are hydrogen-consuming processes but since more than half of the stellar material is hydrogen, the stars are able to maintain their nuclear status quo for an extreme length of time, many times the duration of the contraction period. As the hydrogen of a star is slowly transformed into helium, the star's radius, luminosity, and surface temperature remain essentially unchanged, increasing only slightly when the hydrogen content of the transformation zone drops too low.

But when the hydrogen is gone, the star must resume its long-interrupted contraction and will begin to "walk the last mile" of its life span. Before discussing what goes on in a dying star, we must insert a few remarks about other nuclear reactions.

As the author in collaboration with Edward Teller showed some time ago, some nuclear reactions may be expected to take place at a comparatively early stage of a star's contraction period. If we go over a complete list of all possible nuclear reactions between light elements we find that there are six which have much lower ignition points than either the carbon cycle or the hydrogen-helium process. Those are the reactions between deuterium, lithium, beryllium, and boron on one side, and hydrogen on the other. They are:

$$
\begin{aligned}
&{}_{1}D^{2} + {}_{1}H^{1} \rightarrow {}_{2}He^{3} + \text{radiation}\\
&{}_{3}Li^{6} + {}_{1}H^{1} \rightarrow {}_{2}He^{4} + {}_{2}He^{3}\\
&{}_{3}Li^{7} + {}_{1}H^{1} \rightarrow {}_{2}He^{4} + {}_{2}He^{4}\\
&{}_{4}Be^{9} + {}_{1}H^{1} \rightarrow {}_{3}Li^{6} + {}_{2}He^{4}\\
&{}_{5}B^{10} + {}_{1}H^{1} \rightarrow {}_{6}C^{11} + \text{radiation}\\
&{}_{5}B^{11} + {}_{1}H^{1} \rightarrow {}_{2}He^{4} + {}_{2}He^{4} + {}_{2}He^{4}
\end{aligned}
$$

The ignition points of these six reactions lie between 1 million and 7 million degrees. The fundamental difference between these reactions and the carbon cycle is that they they are not cyclic, which means that the light elements involved in them are not regenerated. Since deuterium, lithium, beryllium, and boron represent only a very small fraction of the total of stellar matter, the interruptions of the contraction period of young stars caused by these nuclear reactions must be comparatively short.

If these short interruptions were plotted on the Hertzsprung-Russell diagram, they would cause a slight accumulation of points along the dotted lines running parallel to the right of the Main Sequence (Fig. 35). But since there are only a very few stars in these regions of the diagram it is difficult to observe accumulations of that kind. Because the main bulk of the stellar population of our galaxy was born and passed through its youthful stages billions of years ago and because the birthrate of new stars in the present era is exceedingly low, it is almost impossible to collect sufficient statistical material for the study of the early contractive stages of stellar evolution. At least no sufficient statistical material has been collected so far.

Aging stars

We now come to the final stage of stellar evolution, the stage when a star approaches exhaustion of its original hydrogen resources. As we have seen in Chapter I, this hydrogen exhaustion may be expected in the present era for all stars of the original stock which have masses up to four or five times that of our sun.

In the Hertzsprung-Russell diagram this belt of aging stars (Fig. 38) intersects the Main Sequence somewhat above Sirius. It is characterized by all kinds of unusual behavior.

To the right of the Main Sequence there is a large group of stars, known as the Red Giants, which possess much larger diameters, much lower mean densities, and considerably lower sur-

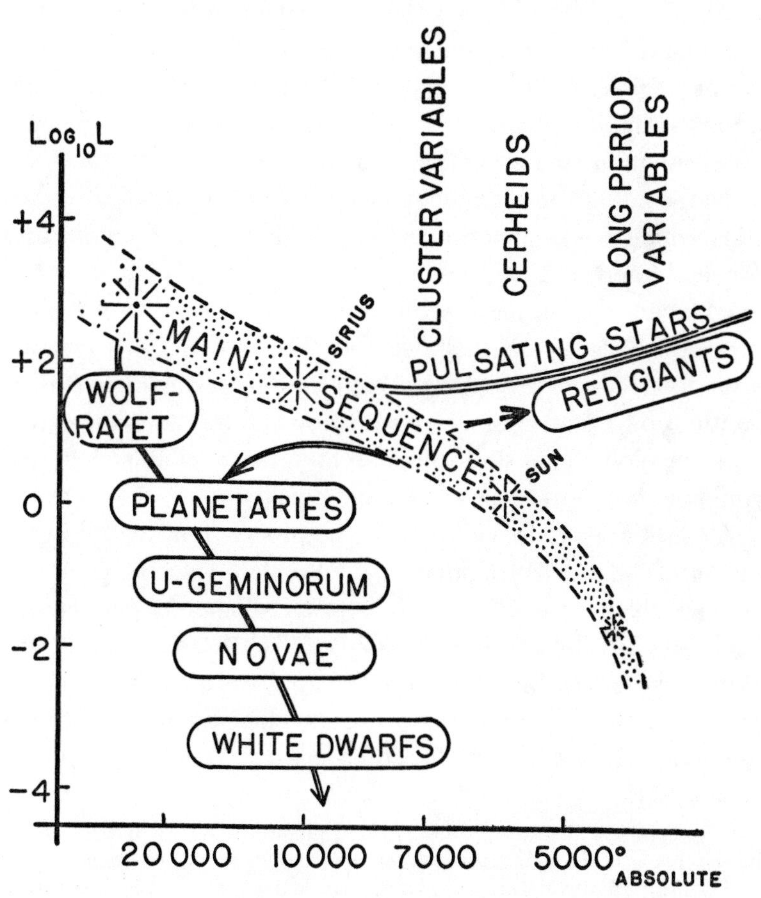

Fig. 38. The belt of aging stars in the Hertzsprung-Russell diagram

face temperatures than the normal Main Sequence stars of the same mass. Many of the stars belonging to this group are apparently in an unstable state and their gigantic bodies are swelling up and subsiding again, causing periodic variations of luminosity. The pulsation period increases as we get farther away from the Main Sequence line. Short pulsation periods, ranging from a few hours to about a day, are characteristic for the so-called "cluster variables," usually found in globular clusters (hence the name) and belonging to Stellar Population II. Cepheid variables pulsate with periods from a few days to several months and are found mostly among Stellar Population I in the regions of the spiral arms. Finally there are the long-period variables which show quite irregular luminosity changes often extending over a period of many years.

To the left of the Main Sequence there is a group of shrunken stars which possess such extremely high surface temperatures (several hundred thousand degrees) that the gas of their outermost layers is continuously blown away by their own radiation pressure. This group includes the Wolf-Rayet stars (named after their two discoverers) which eject gas streams with velocities up to 3000 kilometers per second. It has been estimated that the Wolf-Rayet stars, ejecting their material at such a rate, could not last more than a few million years.

A considerably less violent ejection process is observed in the nuclei of "planetary nebulae." [5] These stars eject gases at the comparatively low velocity of 10 to 20 kilometers per second. The gases are also considerably denser than those ejected by the Wolf-Rayet stars. Being illuminated by the ultraviolet radiation of the hot central star, this gaseous envelope fluoresces, becoming easily accessible to direct observation and photography (Plate VIII).

[5] The name is misleading; there are no planets involved. It was chosen merely because the spherical gaseous envelope of these objects is seen as a disk (see Plate VIII), as planets are when seen through a small telescope.

Located on the same side of the Main Sequence, but somewhat lower in the luminosity scale, are types of stars which are subject to violent periodic explosions. Just as the pulsating stars exhibit a wide range of pulsation periods, so the time intervals between explosions on these stars vary within wide limits. We may first mention the U-Geminorum stars (named after a typical representative of the group), some of which explode every fortnight, while others do so only once in several months. One very important feature of these periodically exploding stars was discovered by Kukarkin and Parenago; namely, that the intensity of the explosions is directly proportionate to the time interval. For example, the star known as AB Draconis explodes on the average every fortnight and during each explosion its brightness increases to fifteen times normal. U-Geminorum itself explodes at time intervals of 97 days with a hundredfold increase in brightness. Thus, the time interval is seven times as long as that of AB Draconis, and the explosion is seven times as powerful. This relationship between time interval and intensity of explosion indicates that in all periodically exploding stars of the U-Geminorum type the average amount of energy released is a constant. The difference between the various representatives is merely one of the intervals of time between releases of the accumulated energy. Hence we are in all probability dealing with stars of identical mass. All thermonuclear reactions are extremely sensitive to stellar mass, which determines the temperature. It would therefore seem impossible for stars of different masses to have the same rate of energy liberation.

If this is true, U-Geminorum stars with different periods may be considered as representing different evolutionary stages in the life of a single star of that type. As we see later, there are indeed good theoretical reasons to expect this kind of behavior from stars that have exhausted their hydrogen supply.

In addition to the U-Geminorum stars there are other periodic explosions of stars, much rarer but much more vigorous. The two stars known as RS Ophiuchi and U Scorpii seem to explode periodically at 30- to 40-year intervals (one repetition has been observed for the first and two for the second). During these explosions their brightness increases several thousand times. Then, of course, there are the ordinary "novae" which increase their luminosity several million times and have never been observed to explode more than once. However, applying the Kukarkin-Parenago relationship, we find that the explosion periods of these stars should be in the neighborhood of 10,000 years, which is much longer than astronomical science has functioned. Plate IX shows a typical luminous gaseous envelope ejected by a nova explosion.

And last, but by no means least, there are the "supernova" explosions, during which a star increases its brightness a billion times, often outshining the whole galaxy to which it belongs. Supernova explosions are very rare indeed. We know thousands of U-Geminorum stars and see a dozen nova explosions in our galaxy every year. But there is only one supernova explosion in four centuries.

It may be that the Star of Bethlehem was the first supernova explosion recorded in the annals of history. However, the first scientifically recorded instance of a supernova was that of the fourth of July (yes!), 1054 A.D. If we look at the point in the sky where contemporary Chinese astronomers located their "new star," we can find a luminous nebulosity, known as the Crab Nebula, with a very faint star in its center (Plate X). Although this object may look similar to one of the planetary nebulae (Plate VIII) or the shell ejected by an ordinary nova (Plate IX), it is actually much bigger. The total mass of the gases forming the Crab Nebula is estimated to be about nine sun masses; the mass of the luminous envelopes of the planetary nebulae never exceeds a few per cent of one sun mass. The velocity of expansion is 1111 kilometers per second, the highest velocity ever observed inside the galaxy. The star in the center of the Crab Nebula which provides the short-wave radiation that makes the expanding gas masses fluoresce seems to be dense and has a surface temperature of about half a million degrees.

Since the Chinese supernova of 1054 only two similar explosions have taken place in our galaxy. One was recorded by the famous Danish astronomer Tycho Brahe in 1572 and the other, in 1604, by Johannes Kepler, Tycho Brahe's pupil and the discoverer of the laws of planetary motion. Although these two explosions must have been every bit as violent as that of 1054 and have presumably ejected similar gaseous envelopes, their remains do not present as spectacular a picture as the Crab Nebula. This is most probably due to unfavorable conditions of illumination.

The extreme rarity of supernovae in our galaxy does not mean that we have to wait for another century or so before being in a position to observe one. If there is one supernova in our own galaxy every 400 years on the average, we can expect to see one supernova per year if we keep several hundred neighboring galaxies under observation. This plan was adopted by Walter Baade and Fritz Zwicky at Mount Wilson Observatory and after a few years they had bagged a sufficiency of excellent supernova pictures providing a multitude of details. The successive stages of supernova development are shown in Plate XI.

After this rapid survey of the various kinds of stellar instability we can now direct our attention to the various attempts at explaining these phenomena in terms of stellar models with dwindling hydrogen content. The present theory of stellar structure holds that the energy liberated by thermonuclear reactions in the center of the stars is transported to the surface by two different processes. In the region immediately surrounding the central source of energy, which contains about 10 per cent of the stellar mass, the temperature gradient is so steep that the gaseous material begins to circulate radially outward, forming

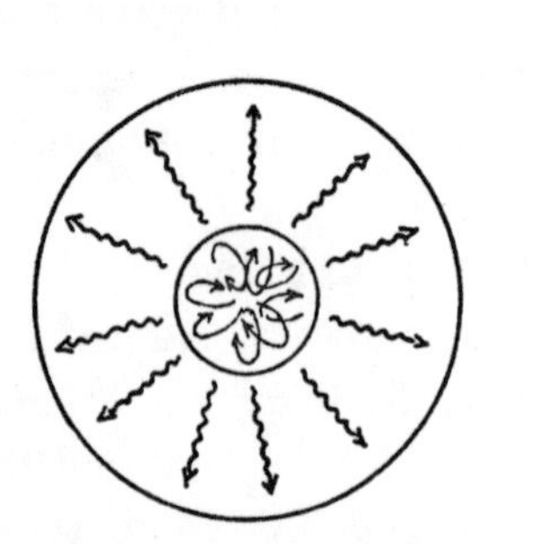

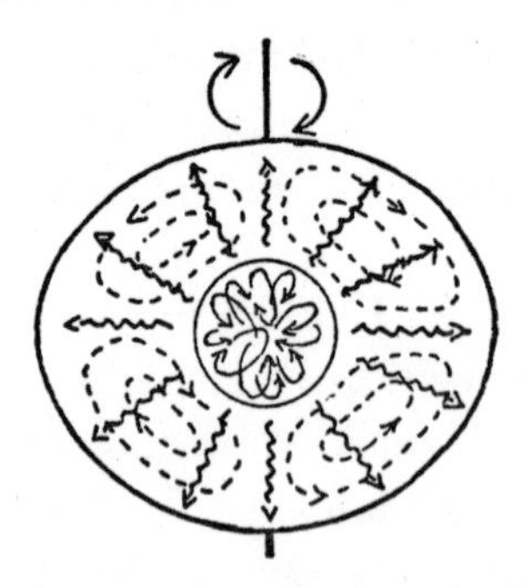

Fig. 39. Convection currents, (left) in a nonrotating star, (right) in a rotating star. The wavy arrows represent the flow of radiation

a multitude of turbulent convection currents. The energy is simply carried outward by fast-moving streams of heated matter, a process similar to that which can be observed in a percolator.

Beyond this convection zone lies the quiet region of the star —the remaining 90 per cent—where there is essentially no motion of material. Heat is transported through these regions by straight conduction, such as carries the heat along a metal bar heated at one end only (Fig. 39, left).

The turbulent currents in the convection zone not only carry the heat, they also mix the material thoroughly, thereby bringing fresh hydrogen into the central region where the nuclear reactions take place. As time goes on, the material in the convection zone gradually becomes dehydrogenized. The outer 90 per cent of the stellar body, through which the heat flows by conduction but in which no thermonuclear reactions take place, may retain its original hydrogen virtually intact.

With rapidly rotating stars, however, the situation is entirely different. According to von Zeipel's theorem, rapidly rotating stars are expected to possess secondary convection currents running from the boundary of the central convection zone all the way to the surface (Fig. 39, right). If this is the case, and if the outer currents are sufficiently fast, the hydrogen content may continue to be uniform throughout the whole body of the star.

If, as may be expected for rapidly rotating stars, the dwindling of the hydrogen supply takes place uniformly throughout the whole body, the final exhaustion of hydrogen content will leave the star in essentially the same state in which it was before the nuclear energy source was switched on. The star is then expected to continue its interrupted contraction. Its position on the Hertzsprung-Russell diagram will then move farther and farther to the left of the Main Sequence, toward the region of lesser radii, higher luminosities, and higher surface temperatures.

At a certain stage of contraction the star's surface temperature is certain to become so high that radiation pressure will begin to tear away its atmosphere; a similar process has been observed in Wolf-Rayet stars and in planetary nebulae. Still later, when the temperature in the central portions of the star reaches several billion degrees, a new and rather unusual nuclear process may be expected to take place. As was shown by the author in collaboration with M. Schönberg, nuclear reactions taking place at such high temperatures can be expected to produce vast quantities of neutrinos,[6] which will escape through the body of the star, carrying away the heat of the interior. The cooling of stellar interiors by this Urca Process [7] is so fast that the entire star can be expected to collapse. In the course of such a sudden collapse very large amounts of enormously hot luminous material would be thrown out. Whether phenomena of this type are actually taking place in the universe and whether they can be held responsible for some of the observed stellar explosions—the supernovae, for example—is still under discussion.

In those cases where the mixing of stellar material is limited

[6] These are nuclear particles ordinarily emitted in the process of β-transformation. Possessing no electric charge and practically no mass, neutrinos have tremendous penetrating power and could be stopped only by a layer of lead several light-years thick!

[7] So named after Casino da Urca in Rio de Janeiro.

to the central convection zone, the evolution of different stars may take radically different courses. It is easy to understand that the complete exhaustion of hydrogen in the central convection zone may result in a spreading of the "nuclear flame" from the center to the former interface between convection and nonconvection regions where hydrogen fuel is still present in large amounts. At that stage the structure of the star would be the so-called "shell-source model" (Fig. 40), consisting of an isothermal dehydrogenized core (a former convection region) sur-

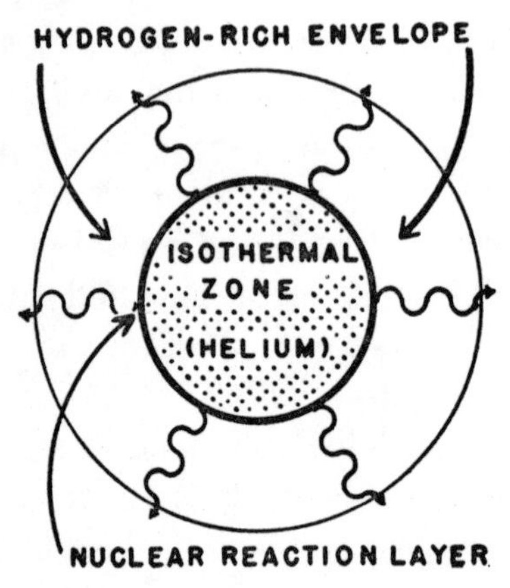

Fig. 40. Shell-source model

rounded by the hydrogen-rich envelope. The nuclear reaction (carbon cycle) will then take place on the inner face of the enveloping shell, gradually eating its way to the surface. It is a process which closely resembles the growing ring of fire in a dry lawn on a still day.

The properties of a shell source model were first studied by the author in collaboration with C. Critchfield (of H–H process fame) and G. Keller and by several other investigators. But much of the probable behavior of such a star is still shrouded in mystery. It seems that it has a choice of two different paths of

Plate X. The Crab Nebula in Taurus

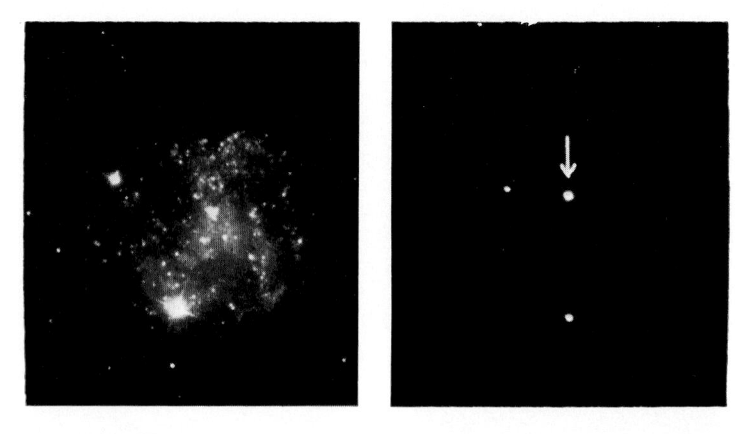

A. Before the explosion; too faint to observe

B. Early in the explosion stage; maximum brightness

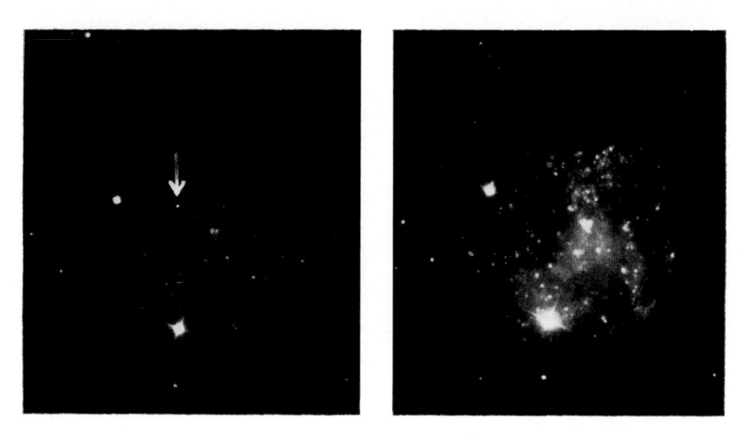

C. Later; becoming faint

D. Still later; too faint to observe

Plate XI. A supernova: four stages
(The photographs shown in A and D were given considerably longer exposures than the others, in order to show the structure of the galaxy)

evolution—both mathematically possible—when the energy-producing shell moves up to the surface.

The first is that the radius of the star may increase almost beyond limit, with a resulting steady drop in its surface temperature. In this case the star will move through the Hertzsprung-Russell diagram toward the region occupied by the Red Giants. A. Reiz in Sweden has actually shown that a model of the star Capella can be constructed on the assumption that carbon-cycle transformation takes place on the inner surface of a shell enclosing a dehydrogenized core.

The second possibility is that the properties of the star will remain quite similar to those of Main Sequence stars until the size of the dehydrogenized core exceeds a certain limit. Then the star will be unable to maintain its normal static equilibrium. The author and C. Longmire have shown that the star can be expected to change to a state in which its radius will be subject to slow periodic changes. When the growing shell source of energy just exceeds the critical value at which equilibrium is possible, these oscillations will have a rather short period and a small amplitude. As the energy-source shell grows, the period grows longer and the amplitude larger.

Of course, the long-period oscillations derived from a mathematical investigation of a shell-source model represent slow changes and do not contribute directly to an understanding of the spontaneous explosion process. However, at certain stages the boundary between the hot dehydrogenized core and the somewhat cooler hydrogen-rich envelope may develop a convectional instability, and a fast mixing of the two regions take place. If that happened we would get an effect similar to that obtained by pouring a bucket of cold gasoline on red-hot coals: nuclear energy would be released in a flash and the star would blow up like a bomb. The magnitude of the explosion will naturally depend on the mass of the unstable boundary region.

It is likely to be larger the farther the energy-producing shell had progressed toward the surface. If this is correct we would have a star model characterized by periodic explosions of ever-increasing intensity, occurring at ever-lengthening intervals. This might be the explanation of the U-Geminorum → nova → supernova sequence described earlier.

If we accept this point of view we have to consider all stellar explosions as successive stages in the evolution of an aging star, beginning with the minor puffs of the U-Geminorum type and finishing with the fantastic blow-up of a supernova. But this theory is at present based largely on intuition and guesswork. It cannot be considered as established until a mathematical analysis, lengthy and tedious by its very nature, has actually been completed. Such calculations are now being contemplated and will probably be carried out by means of an electronic computer. One can get a good idea of the complexity of this work from the rather conservative estimate that the solution would require the services of one hundred human computers for at least one hundred years. Modern electric computers, however, can accomplish such calculations within a reasonable time, and it may be hoped that the problem will be solved in the near future.

Before leaving the problem of exploding stars we must point out that the comparison of various theoretical possibilities with existing observational material is complicated by the fact that the listing of known stellar explosions might lump together stars of two different age groups. On the one hand, there are the aging stars of the original stock, weighing four or five times as much as our sun, and having lived for about 3 billion years before their hydrogen supply was exhausted. On the other hand, there are the much more massive stars of much more recent origin which burned their hydrogen too fast. Before we can try to correlate observed events and base a theory on the observations and their probable correlation, we must be able to separate the two groups, and that unfortunately is by no means an easy task.

Dead stars

So far we have dealt with the whole life of a star, from original condensation to spectacular outbreaks which may be compared to the convulsions of death agony.

But how does a dead star look?

If we define a dead star as one in which no sources of energy are left and consequently no further evolutionary changes are possible, the stars known as White Dwarfs are dead. This is a group of extremely dense and hot stars located in the lower left-hand corner of the Hertzsprung-Russell diagram (Fig. 35).

The so-called Companion of Sirius is a typical example of a White Dwarf. Its diameter is not much larger than that of the earth but its mass is about the same as that of our sun. White Dwarfs represent the extreme stage of stellar contraction and their mean density is about a million times that of water. Theoretical study of their internal structure indicates that they have lost the last trace of hydrogen and are still hot only because they have not yet had enough time to cool off and to become dark bodies. Their warmth is that of a corpse a few minutes after death. But because of their enormous heat content, it will take them billions of years to cool.

In our galaxy there is approximately one White Dwarf to ten actively living stars of the Main Sequence, indicating that our universe is still comparatively young and its cemeteries are not yet overcrowded. Most probably all the White Dwarfs in our galaxy are the remains of rather massive stars which used up their hydrogen supply in a wasteful manner. The faint star in the center of the Crab Nebula is in all probability one of the most recent cases of stellar mortality. Of course, it is not yet as dense as the Companion of Sirius, and is also still considerably hotter than that star, which must have been dead for several hundred million years. But it will reach that state eventually, and, as time goes on, there will be more and more stellar corpses in the vast expanses of the universe.

Conclusion

We now come to the end of our discourse, and a picture of the creative process begins to emerge—somewhat hazy and fragmentary but in its general outlines quite definite. In the dim pregalactic past we perceive a glimpse of a metaphysical "St. Augustine's Era" when the universe, whatever it was made of, was involved in a gigantic collapse. Of course, we have no information about that era, which could have lasted from the minus infinity of time to about five billion years ago, since all "archaeological records" pertaining to that distant past must have been completely obliterated when the cosmic masses were squeezed into a pulp. The masses of the universe must have emerged from the Big Squeeze in a completely broken-up state, forming the primordial Ylem of neutrons, protons, and electrons. As the Ylem cooled rapidly through expansion, these elementary particles began to stick to one another, forming aggregates of different complexities which were the prototypes of the atomic nuclei of today. During this early period of "nuclear cooking," which lasted not more than an hour of time, conditions throughout the universe closely approximated those existing in the center of an exploding atomic bomb. Cosmic space was full of high-energy gamma radiation, the mass-density of which greatly exceeded the density of ordinary atomic matter. The temperature throughout the universe was in the neighborhood of a billion degrees, but the density of matter was comparable to the density of atmospheric air at high altitudes.

Following that highly productive first hour of the history of our universe, nothing in particular happened for the next 30 million years. The gas, consisting of the newly formed atoms, continued to expand, and its temperature became lower and lower. Radiant energy, which at the beginning played a predominant role in the evolutionary process, gradually lost its importance and by the end of the thirty-millionth year yielded its priority in favor of ordinary atomic matter. As soon as matter took over, the force of Newtonian gravity, which represents one of the most important characteristics of "ponderable" matter, came into play, breaking up the hitherto homogeneous gas into gigantic clouds, the proto-galaxies. In that era the temperature dropped to approximately that which we call "room temperature," so that space was still rather warm, although completely dark.

While the original proto-galaxies were being driven farther and farther apart by continued expansion, material in their interiors began to condense into a multitude of much smaller aggregations, called proto-stars. Because of the comparatively small size of these proto-stars their contraction progressed quite rapidly. Very soon the temperature in their interiors reached the value at which nuclear reactions between hydrogen and various light elements would take place, and space became bright again, being illuminated by myriads of stars. When the stars were formed by the condensation of the gaseous material of the proto-galaxies, some of that material was left over in their vicinity and from it sprang planetary systems. The planets were too small to create their own sources of nuclear energy; they cooled off fast and developed solid rocky crusts. With the help of the radiations from their respective suns, certain chemical compounds which were present on the surfaces of these planets went through an evolutionary process, as yet not well understood, by which organic materials of higher and higher complexity were developed. Thus the naked rocky surfaces of the planets were presently covered by the green carpets of woods and meadows. Animals appeared, first primitive and then more and more complicated, finally evolving into the human being who is intelligent enough to ask and to answer questions concerning the events which took place billions of years before he came into existence.

Probably one of the most striking conclusions from our inquiry into the history of the universe is the fact that the main evolutionary events of physical development occupied only such a tiny fraction of the total period. This, of course, only means that organic evolution takes place at a much slower rate than the large-scale physical processes in the universe.

Indeed, it took less than an hour to make the atoms, a few hundred million years to make the stars and planets, but five billion years to make man!

Appendix

Mathematics of Holmes' Method (Addendum to Chapter I)

Assuming that geochemical processes can change relative concentrations of uranium and lead in various samples but never change the relative abundance of the isotopes, Holmes gives the following three equations for three samples of different geological ages t_1, t_2, and t_3:

$$\frac{a_i - x}{b_i - y} = R(t_0)\,\frac{1 - e^{\lambda(t_0 - t_i)}}{1 - e^{\lambda'(t_0 - t_i)}} \quad (i = 1, 2, 3)$$

where a_i and b_i are observed relative concentrations of Pb^{206} and Pb^{207} (in respect to Pb^{204}) in the sample for the age t_i; λ and λ', the decay constants of U^{238} and U^{235}; $R(t_0)$ the relative amount of U^{238} and U^{235} at the zero age t_0 (calculated from their present observed relative amounts and their decay constants); and x and y relative concentrations of primeval Pb^{206} and Pb^{207} (in respect to Pb^{204}). Solving the equations for different triplets of samples, he obtains in each case the values of x, y, and t_0. The quoted value $t_0 = 3.35 \cdot 10^9$ yr is the mean value obtained from many triplets of samples.

Kinetic and Potential Energy of Expansion (Addendum to Chapter II)

Consider a large sphere of radius R, containing so many galaxies that we can assume it as being uniformly filled with matter (Fig. 3, right). The matter is moving away from the center of that sphere, which we might as well imagine as coinciding with our own galaxy. The velocity of the outward movement is the higher the farther the matter is from the center. According to Hubble's law, galaxies located at the distance r are moving with the velocity $v = \alpha \cdot r$, where we have written α for the numerical coefficient in Hubble's formula given in the text. The kinetic energy of the moving matter inside the sphere is given, as usual, by the product of mass multiplied by the square of velocity and then divided by 2. Since the velocity changes from the value 0 near the center to the value $\dot{R} = \alpha \cdot R$ at the periphery, we have to use some average value. Mathematical analysis shows that we must use three-fifths of the peripheral velocity. Using this figure, and writing ρ for the density of matter in the sphere, we obtain for the kinetic energy K the expression

$$K = 1.26\rho R^3 \dot{R}^2 = 1.26\rho\alpha^2 R^5 \qquad (1)$$ [1]

On the other hand, the potential energy U of Newtonian attraction between the masses within the sphere is

$$U = -7 \cdot 10^{-7} \cdot \rho^2 R^5 \qquad (2)$$ [2]

Since both K and U are proportional to the same power of R (the fifth power), their ratio is independent of the size of the sphere under consideration. We obtain:

$$\frac{K}{U} = \frac{1.8 \cdot 10^6 \cdot \alpha^2}{\rho} \qquad (3)$$

Taking $\alpha = 1.9 \cdot 10^{-17}$, and accepting for the mean density of matter in the universe the most probable value $\rho = 10^{-30}$ g/cm^3

[1] In expanded form:

$$K = \tfrac{1}{2}(\tfrac{4}{3}\pi R^3 \rho)^2 \cdot \tfrac{3}{5}(\alpha \dot{R})^2 = \tfrac{2}{5}\pi\rho\alpha^2 R^5$$

[2] In expanded form:

$$U = G/RM^2 = -G/R(\tfrac{4}{3}\pi\rho R^3)^2 = -\tfrac{16}{15}\pi^2 G\rho^2 R^5$$

where Newton's constant is

$$G = 6.66 \cdot 10^{-8}$$

(one hydrogen atom per cubic meter) we find that the ratio K/U is 650. Obviously the kinetic energy of the expanding universe is much greater than the potential energy of Newtonian gravity. The galaxies are flying away from one another with velocities much higher than the corresponding escape velocity, and *the expansion will never stop.* If, with Baade, we assume that intergalactic distances are 2½ times as large as previously assumed, a will be one-half its previous value, and the value for the mean density of the universe will be one-fifteenth of that used. Then formula (3) will give $K/U = 1650$, and our conclusion will still hold with an even larger "margin of safety."

Temperature and Density Variations in an Expanding Universe (Addendum to Chapter II)

Using expressions (1) and (2) for the kinetic and potential energies in an expanding universe we can now write the equation governing that expansion by stating that the sum of kinetic and potential energies in any part of the universe must remain constant throughout the expansion (mechanical law of conservation of energy). Thus we obtain:

$$1.26\rho\dot{R}^2R^3 - 7 \cdot 10^{-7}\rho^2R^5 = \text{constant} \qquad (4)$$ [3]

which gives us the relation between any distance R and its rate of change $\dot{R}$. This equation, derived by simple mechanical considerations, is exactly the same that one would obtain from the complicated tensor equations of Einstein's general theory of relativity. We now want to learn the consequences of these equations for the early stages of expansion when both the density and temperature in the universe were exceedingly high.

The recognition of the great importance of radiation during the early stages of the expanding universe permits us to derive

[3] In expanded form:

$$\tfrac{5}{2}\pi\rho\dot{R}^2R^3 - \tfrac{15}{16}\pi^2 G\rho^2R^5 = \text{constant}$$

a simple expression for the temperature of space at different periods after the beginning of expansion. Considering ρ in equation (4) to represent exclusively the density of radiation, we find [4] that t seconds after the beginning of expansion the temperature of space must have been:

$$T = \frac{1.5 \cdot 10^{10}}{\sqrt{t}} \text{ degrees absolute} \qquad (5)$$

While the theory provides an exact expression for the temperature in the expanding universe, it leads to an expression with an unknown factor for the density of matter. One can easily find [5] that the density of matter in the universe must have been dependent on its age, according to the formula:

$$\rho = \frac{\rho_0}{t^{3/2}} \qquad (6)$$

which, however, contains ρ_0 as the unknown constant.

The Building-up Process (Addendum to Chapter III)

Mathematically, the equation of the building-up process can be written as

$$\dot{n}_A = vn_0 \left(\sigma_{A-1}\, n_{A-1} - \sigma_A n_A\right) \qquad (7)$$

where n_0 is the number of neutrons, n_A and n_{A-1} the number of

[4] The equation in footnote 3 can be rewritten as

$$\frac{\dot{R}}{R} = \sqrt{\frac{8\pi\, G\rho}{3} + \frac{\text{constant}}{R^2}}$$

Assuming that $\rho = aT^4/c^2$ (Stephan-Boltzmann law) and $\dot{R}/R = -\dot{T}/T$ (Wien law) and neglecting the second term under the radical, we get:

$$-\dot{T}/T = \sqrt{\frac{8\pi G}{3} \cdot \frac{aT^4}{c^2}}$$

which results in:

$$T = \sqrt[4]{\frac{3c^2}{32\,\pi\, Ga}} \cdot \frac{1}{t^{\frac{1}{2}}} = \frac{1.5 \cdot 10^{10}}{t^{\frac{1}{2}}}$$

[5] We have $\rho \sim 1/R^3 \sim T^3 \sim 1/t^{\frac{3}{2}}$

complex nuclei with weights A and $A-1$ respectively, and v the velocity of neutrons determining the number of collisions. Since the calculations are made for expanding and cooling Ylem, temperature and density of the material must be assumed to depend on time, in accordance with the preceding formulae.

Index

(*Figures in italics refer to illustrations, figures followed by an n to footnotes.*)